TK 7881.8 .B37 1991

Bartlett, Bruce.

Stereo microphone techniques

NEW ENGLAND INSTITUTE
OF TECHNOLOGY
LEARNING RESOURCES CENTER

STEREO MICROPHONE TECHNIQUES

BRUCE BARTLETT

FOCAL PRESS
Boston   London

CONTENTS

Preface   xiii

Acknowledgments   xvii

1. **Microphone Polar Patterns and Other Specifications**   1
   Polar Patterns   1
     Advantages of Each Pattern / 3
     Other Polar-Pattern Considerations / 3
   Transducer Type   5
   Maximum Sound Pressure Level   7
   Sensitivity   7
   Self-Noise   8
   Signal-to-Noise Ratio (S/N)   8
   Microphone Types   9
     Free-Field Microphone / 9
     Boundary Microphone / 9
     Stereo Microphone / 10
     Shotgun Microphone / 11
     Parabolic Microphone / 11
   Microphone Accessories   11
     Stands and Booms / 11
     Stereo Microphone Adapter / 12
     Shock Mount / 12
     Phantom-Power Supply / 13
     Junction Box and Snake / 13
     Splitter / 13
   Reference   13

2. **Overview of Stereo Microphone Techniques**   14
   Why Record in Stereo?   14
   Other Applications for Stereo Miking   15
   Goals of Stereo Miking   16

Types of Stereo Microphone Techniques — 18
   Coincident Pair / 18
   Spaced Pair / 21
   Near-Coincident Pair / 23
Comparing the Three Stereo Miking Techniques — 24
Mounting Hardware — 25
Microphone Requirements — 25
References — 26

3. Stereo Imaging Theory — 27

Definitions — 27
How We Localize Real Sound Sources — 29
How We Localize Images between Speakers — 32
Requirements for Natural Imaging over Loudspeakers — 34
Currently Used Image-Localization Mechanisms — 38
   Localization by Amplitude Differences / 39
   Localization by Time Differences / 41
   Localization by Amplitude and Time Differences / 43
   Summary / 43
Predicting Image Locations — 44
Choosing Angling and Spacing — 48
Spaciousness and Spatial Equalization — 49
References — 50

4. Specific Free-Field Stereo Microphone Techniques — 53

Localization Accuracy — 53
Examples of Coincident-Pair Techniques — 56
   Coincident Cardioids Angled 180° Apart / 56
   Coincident Cardioids Angled 120° to 135° Apart / 57
   Coincident Cardioids Angled 90° Apart / 57
   Blumlein or Stereosonic Technique (Coincident Bidirectionals Angled 90° Apart) / 57
   Hypercardioids Angled 110° Apart / 59
   XY Shotgun Microphones / 60
Examples of Near-Coincident Techniques — 60
   ORTF System: Cardioids Angled 110° Apart and Spaced 17cm (6.7") Horizontally / 60
   NOS System: Cardioids Angled 90° Apart and Spaced 30cm (12") Horizontally / 62
   OSS (Optimal Stereo Signal or Jecklin disk) / 63
Examples of Spaced-Pair Techniques — 64
   Omnis Spaced 3 Feet Apart / 65
   Omnis Spaced 10 Feet Apart / 66

Three Omnis Spaced 5 Feet Apart (10 Feet End-to-End) / 66
Other Coincident-Pair Techniques 67
MS (Mid-Side) / 67
Calrec Soundfield Microphone / 73
Coincident Systems with Spatial Equalization (Shuffler Circuit) / 73
Other Near-Coincident Techniques 75
Stereo 180 System / 75
Faulkner Phased-Array System / 76
Near-Coincident Systems with Spatial Equalization / 76
Near-Coincident/Spaced-Pair Hybrid / 77
Comparisons of Various Techniques 77
Michael Williams, "Unified Theory of Microphone Systems for Stereophonic Sound Recording" / 77
Carl Ceoen, "Comparative Stereophonic Listening Tests" / 78
Benjamin Bernfeld and Bennett Smith, "Computer-Aided Model of Stereophonic Systems" / 79
C. Huggonet and J. Jouhaneau, "Comparative Spatial Transfer Function of Six Different Stereophonic Systems" / 80
M. Hibbing, "XY and MS Microphone Techniques in Comparison" / 80
Summary / 81
References 81

5. **Stereo Boundary-Microphone Arrays** 83

Floor-Mounted Techniques 83
Floor-Mounted Boundary Microphones Spaced 4 Feet Apart / 84
Floor-Mounted Directional Boundary Microphones / 84
L-Squared Floor Array / 85
OSS Boundary Microphone Floor Array / 87
Floor-Mounted Boundary Microphone Configured for MS / 87
Raised-Boundary Methods 87
PZM® Wedge / 88
L-Squared Array / 89
Pillon PZM® Stereo Shotgun / 89
The Stereo Ambient Sampling System™ / 90
References 94

6. Binaural and Transaural Techniques — 96

Binaural Recording and the Artificial Head — 96
   How It Works / 97
   In-Head Localization / 99
   Artificial-Head Equalization / 100
   Artificial-Head Imaging with Loudspeakers / 100
Transaural Stereo — 101
   How Transaural Stereo Works / 101
   History of Transaural Stereo / 104
   Cooper and Bauck's Crosstalk Canceller / 104
   Lexicon's Transaural Processor / 105
   Other Surround-Sound Systems / 106
References — 107

7. Stereo Recording Procedures — 111

Equipment — 111
Choosing the Recording Site — 113
Session Setup — 113
   Monitoring / 116
Microphone Placement — 116
   Miking Distance / 116
   Stereo-Spread Control / 117
   Soloist Pickup and Spot Microphones / 120
   Setting Levels / 122
   Multitrack Recording / 122
Stereo Miking for Pop Music — 122
Troubleshooting Stereo Sound — 124
   Distortion in the Microphone Signal / 124
   Too Dead (Insufficient Ambience, Hall Reverberation, or Room Acoustics) / 124
   Too Detailed, too Close, too Edgy / 124
   Too Distant (too much Reverberation) / 125
   Narrow Stereo Spread / 125
   Excessive Separation or Hole-in-the-Middle / 126
   Poorly Focused Images / 126
   Images Shifted to One Side (Left-Right Balance Is Faulty) / 126
   Lacks Depth (Lacks a Sense of Nearness and Farness of Various Instruments) / 127
   Lacks Spaciousness / 127
   Early Reflections too Loud / 127
   Bad Balance (Some Instruments too Loud or too Soft) / 127

Contents ☐ xi

    Muddy Bass / 127
    Rumble from Air Conditioning, Trucks,
       and so on / 128
    Bad Tonal Balance (too Dull, too Bright,
       Colored) / 128
  References   128

8. Broadcast, Film and Video, Sound Effects, and Sampling   **130**

  Stereo Television   131
    Imaging Considerations / 131
    Mono-Compatibility / 132
    Monitoring / 132
    Electronic News Gathering (E.N.G.) / 133
    Audience Reaction / 134
    Parades / 136
    Sports / 138
  Stereo Radio   139
    Radio Group Discussions / 139
    Radio Plays / 140
  Film and Video   141
    Feature Films / 141
    Documentaries and Industrial Productions / 142
  Sound Effects   142
  Sampling   143
  References   144

9. Stereo Microphones and Accessories   **146**

  Stereo Microphones   146
  Dummy Heads   153
  Stereo Microphone Adapters (Stereo Bars)   153
  Matrix Decoders   155
  Company Addresses   157

Glossary   **161**

Index   **173**

PREFACE

True-stereo microphone techniques use two or three microphones to capture the overall sound of a sonic event. The stereo recording made from these microphones is usually reproduced over two speakers. Ideally, the goal is to produce a believable illusion of the original performance and its acoustic environment in a solid, or three-dimensional, way.

There are many ways to make true-stereo recordings, and this book, *Stereo Microphone Techniques*, covers them all. The book was written for recording engineers, broadcasters, film-sound engineers, and audio hobbyists. It offers a clear, practical explanation of stereo miking theory, along with specific techniques, procedures, and hardware.

Stereo miking has several applications:

- Classical-music recording—ensembles and soloists
- Pop-music recording—stereo pickup of piano, drums, percussion, and backing vocals
- Stereo TV—talk shows, game shows, audience reaction, Electronic News Gathering, sports, parades
- Stereo film—feature film dialog and ambience, documentaries
- Stereo radio—group discussions, radio plays
- Stereo sampling and sound effects

For example, an orchestra might be recorded with two microphones and played back over two speakers. You would hear sonic images of the instruments in various locations between the stereo pair of speakers. These image locations—left to right, front to back—correspond to the instrument locations during the recording session. In addition, the concert hall acoustics are reproduced with a pleasing spaciousness. The result can be a beautiful, realistic recreation of the original event.

The book starts with an explanation of microphone polar patterns, which is necessary to understand how stereo techniques work. This is followed by a simple overview of the most common stereo microphone techniques.

Next, stereo imaging theory is covered in detail: how we localize real sound sources, how we localize images between loudspeakers, and how stereo microphone techniques create images in various locations. You'll learn how to configure stereo arrays to achieve specific stereo effects. Spaciousness and spatial equalization are covered as well.

Specific microphone techniques—such as MS (mid-side), Blumlein, ORTF, OSS, SASS—are explained next: their characteristics, stereo effects, benefits, and drawbacks. One chapter is devoted to free-field methods, another to boundary methods. The latest techniques are explained in detail, such as current developments in binaural and transaural stereo.

Armed with this knowledge, you're ready to record a musical ensemble. The necessary step-by-step procedures are described. A troubleshooting guide helps you pinpoint and solve stereo-related problems. Other stereo applications are explored as well: television, video, film, sound effects, and sampling.

The book concludes with a current listing of stereo microphones and accessories, and a glossary. Each chapter includes a list of references for further reading.

This is the first textbook ever written on the subject, even though stereo miking has a history going back to 1881. It was first demonstrated at the International Exhibition of Electricity in Paris. A performance of the Paris Opera was picked up with several spaced pairs of microphones and transmitted 3 km away to people listening binaurally with two earphones [1].

In the 1930s, British researcher Alan Blumlein and Bell Labs engineers Arthur Keller and Harvey Fletcher independently developed stereo technology for disk reproduction [2,3]. Blumlein's patent in particular is a classic, far ahead of its time, and should be required reading for audio engineers. Stereo recording and reproduction became standard practice when stereo LPs were introduced in the late 1950s.

Our demands on stereo listening have become more sophisticated over the years. At first we were excited to hear simple left-right (ping-pong) stereo. Next we added a center image. Then we wanted accurate localization at all points, sharp imaging, and depth. Currently, methods are being devised to reproduce sound images all around the listener with only two loudspeakers in front.

The need for up-to-date information on stereo microphone techniques has never been greater, and I hope this book answers that need.

REFERENCES

1. B. Hertz, "100 Years with Stereo: The Beginning," *Journal of the Audio Engineering Society*, Vol. 29, No. 5, 1981 (May), pp. 368–372.
2. A. Blumlein, British Patent Specification 394,325, *J. Audio Eng. Soc.*, Vol. 6, No. 2, 1958 (April), p. 91.
3. A. Keller, "Early Hi-Fi and Stereo Recording at Bell Laboratories (1931–1932)," *J. Audio Eng. Soc.*, Vol. 29, No. 4, 1981 (April), pp. 274–280.

All the above references can be found in *Stereophonic Techniques*, an anthology published by the Audio Engineering Society, 60 E. 42nd St., New York, NY 10165.

ACKNOWLEDGMENTS

Thank you to all the microphone manufacturers who sent photographs for this book. Thanks also to Terry Skelton, Mike Billingsley, Dan Gibson, and Ed Kelly for their suggestions on stereo miking.

I greatly appreciate the contributions and advice of reviewers Jim Loomis, an instructor at Ithaca College; Bruce Outwin, an instructor at Emerson College; and Ron Estes of NBC in Burbank. Thanks also to Phil Sutherland, Associate Editor at Focal Press, and Claire Huismann, Production Coordinator at Publication Services.

Thank you to the publishers who allowed me to use some of my own material for this book. Chapters 1 and 6 are reprinted with permission from

MR&M Publishing Corp. and Sagamore Publishing Co. Inc., "Recording Techniques" series by Bruce Bartlett.

Howard W. Sams & Co., *Introduction to Professional Recording Techniques*, Chapters 7 and 17, copyright 1987 by Bruce Bartlett.

Radio World, "Stereo Microphone Techniques Part 1," Nov. 1989, and "Stereo Microphone Techniques Part 2," Feb. 1990, by Bruce Bartlett.

Parts of Chapters 3 and 5 were based on the article by B. Bartlett, "An Improved Stereo Microphone Array Using Boundary Technology: Theoretical Aspects," *Journal of the Audio Engineering Society*, Vol. 38, No. 7/8, 1990 (July/August), pp. 543–552.

STEREO MICROPHONE TECHNIQUES

1

Microphone Polar Patterns and Other Specifications

☐

Before we can understand how stereo microphone arrays work, we need to understand microphone polar patterns. These are explained in this chapter, as are other specifications that will help you choose appropriate microphones and accessories for stereo recording.

POLAR PATTERNS

Microphones differ in the way they respond to sounds coming from different directions. Some respond the same to sounds from all directions; others have different output levels for sources at different angles around the microphone.

This varying sensitivity versus angle can be graphed as a polar pattern or polar response (see Figure 1-1). Polar patterns are plotted on polar graph paper as follows: In an anechoic chamber, the microphone is exposed to a tone of a single frequency and its output voltage is measured as it is rotated around its diaphragm. The voltage at 0° (on-axis) is called the "0 dB reference," and the voltages at other angles are referenced to that. In other words, the polar-response graph plots relative sensitivity in dB versus angle of sound incidence in degrees. Often, several such plots are made at various frequencies. Note that the microphone rotation should be clockwise if the degree marks increase in a counter-clockwise direction.

Another way to generate a polar plot is by using time delay spectrometry. Measure the microphone's frequency response every 10° around the microphone, then process the data with a program that draws a polar plot at selected frequencies.

The three major polar patterns are omnidirectional, unidirectional, and bidirectional. An omnidirectional microphone is equally sensitive to sounds arriving from all directions. A unidirectional microphone

2 □ STEREO MICROPHONE TECHNIQUES

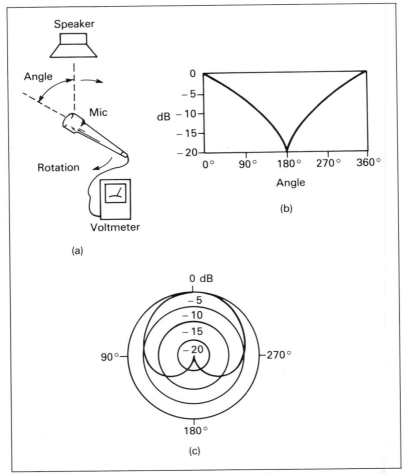

Figure 1-1 Plotting microphone polar responses: (a) measurement setup; (b) mic output versus angle, plotted on rectangular coordinates; (c) mic output versus angle, plotted on polar coordinates.

is most sensitive to sounds arriving from one direction—in front of the microphone—therefore discriminating against sounds entering the sides or rear of the microphone. A bidirectional microphone is most sensitive to sounds arriving from two directions—in front of and behind the microphone—but rejects sounds entering the sides.

The unidirectional classification can be further divided into cardioid, supercardioid, and hypercardioid pickup characteristics. A microphone with a cardioid pattern is sensitive to sounds arriving from a broad angle in front of the microphone. It is about 6 dB less sensitive at the sides and about 15 to 25 dB less sensitive at the rear. To hear how a cardioid pickup pattern works, talk into a cardioid microphone from all sides while listening to its output. Your reproduced voice will be loudest when you talk into the front of the microphone and softest when you talk into the rear.

The supercardioid pattern is 8.7 dB down at the sides and has two *nulls*—points of least pickup—at ±125° off axis. "Off-axis" means "away from the front." The hypercardioid pattern is 12 dB down at the sides and has two nulls of least pickup at 110° either side off axis.

Figure 1-2 shows various polar patterns. Note that a polar plot is not a geographical map of the "reach" of a microphone; a microphone does not suddenly become dead outside its polar pattern. There is no "outside." The graph merely plots sensitivity at one frequency as distance from the origin; it is not the spatial spread of the pattern.

☐ Advantages of Each Pattern

Omnidirectional microphones have several characteristics that make them especially useful for certain applications. Use omnidirectional microphones when you need

- all-around pickup
- extra pickup of room reverberation
- low pickup of mechanical vibration and wind noise
- extended low-frequency response (in condenser mics)
- lower cost in general
- freedom from proximity effect (up-close bass boost)

Use directional microphones when you need

- rejection of room acoustics and background noise
- coincident or near-coincident stereo (explained in the next chapter)

☐ Other Polar-Pattern Considerations

In most microphones, it's desirable that the polar pattern stay reasonably consistent at all frequencies. If not, you'll hear off-axis coloration: the mic will sound tonally different on and off axis. Uniform polar patterns at different frequencies indicate similar frequency responses at different angles of incidence.

Some condenser mics come with switchable polar patterns.

An omnidirectional boundary microphone (a surface-mounted microphone, explained later) has a half-omni, or hemispherical, polar pattern. A unidirectional boundary microphone has a half-supercardioid or half-cardioid polar pattern. The boundary mounting increases the directionality of the microphone, thus reducing pickup of room acoustics.

4 ☐ STEREO MICROPHONE TECHNIQUES

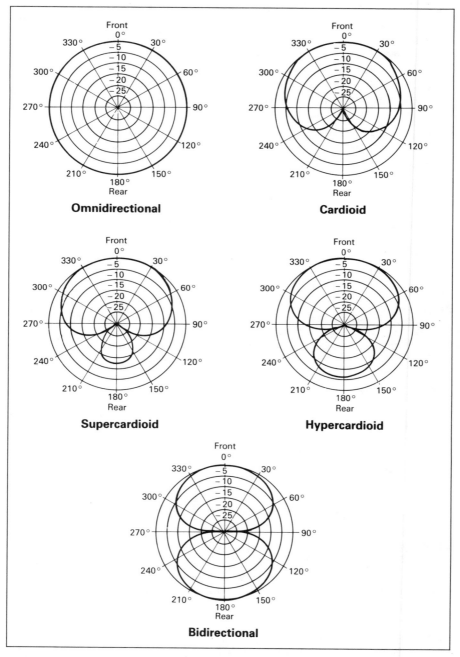

Figure 1-2 Various polar patterns. Sensitivity is plotted versus angle of sound incidence.

TRANSDUCER TYPE

We've seen that microphones differ in their polar patterns. They also differ in the way they convert sound to electricity. The three operating principles of recording microphones are condenser, moving coil, and ribbon.

In a condenser microphone (shown in Figure 1-3), a conductive diaphragm and an adjacent metallic disk (called a backplate) are charged with static electricity to form a capacitor. Sound waves vibrate the diaphragm; this vibration varies the capacitance and produces a voltage similar to the incoming sound wave. The diaphragm and backplate can be charged (biased) by an external power supply, or they can be permanently charged by an electret material in the diaphragm or on the backplate. *Capacitor* is the modern term for "condenser," but the name *condenser microphone* has stuck and is popular usage. Some condenser mics use variable capacitance to frequency-modulate a radio-frequency carrier.

The condenser microphone is the preferred type for stereo recording because it generally has a wide, smooth frequency response, a detailed sound quality, and high sensitivity. It requires a power supply to operate, such as a battery or phantom power. Phantom power is provided by an external supply or from the recording mixer.

In a moving-coil microphone (sometimes called "dynamic"), a voice coil attached to a diaphragm is suspended in a magnetic field (as shown in Figure 1-4). Sound waves vibrate the diaphragm and voice coil, pro-

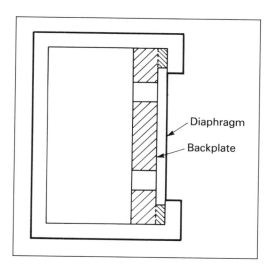

Figure 1-3 Condenser microphone.

Figure 1-4 Moving-coil microphone.

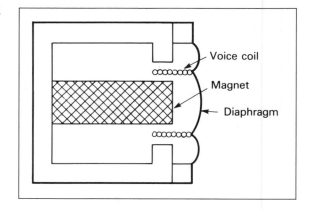

ducing an electric signal similar to the incoming sound wave. No power supply is needed.

In a ribbon microphone (Figure 1-5), the conductor is a thin metallic strip, or ribbon. Sound waves vibrate the ribbon, which generates electricity. A printed ribbon microphone has a thin plastic diaphragm with an implanted ribbon.

Since the moving-coil microphone generally has a rougher frequency response and lower sensitivity than the condenser type, the moving-coil type is less often used for stereo recording. The ribbon, however,

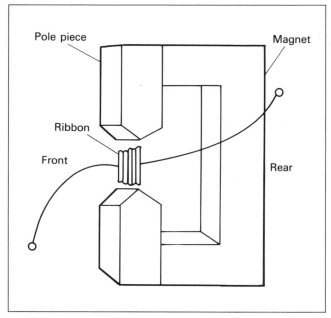

Figure 1-5 Ribbon microphone.

has a smooth response and has been used in Blumlein arrays (crossed bidirectionals). The printed-ribbon type is currently used in two stereo microphones, but most stereo mics are condensers.

MAXIMUM SOUND PRESSURE LEVEL

Another microphone specification is maximum sound pressure level (SPL). To clarify this specification, first we need to explain the term *SPL*. SPL is a measure of the intensity of a sound. The quietest sound we can hear, the threshold of hearing, measures 0 dB SPL. Normal conversation at a distance of one foot measures about 70 dB SPL; painfully loud sound is above 120 dB SPL.

Maximum SPL is the sound pressure level at which a microphone's output signal starts to distort; this is usually the SPL at which the microphone produces 3 percent total harmonic distortion (THD). (Some manufacturers use 1 percent THD.) If a microphone has a maximum SPL spec of 125 dB SPL, that means the microphone starts to distort audibly when the sound pressure level produced by the source reaches 125 dB SPL. A maximum SPL spec of 120 dB is good, 135 dB is very good, and 150 dB is excellent.

SENSITIVITY

Sensitivity is another specification to consider. It is a measure of the efficiency of a microphone. A very sensitive microphone produces a relatively high output voltage for a sound source of a given loudness.

Microphone sensitivity is often stated in "dB re 1 volt (dBV) per microbar." This figure tells what voltage the microphone produces (in dB relative to 1 volt) when picking up a 1000-Hz tone at a 74 dB SPL. Sensitivity also might be stated in millivolts/Pa, where 1 Pa = 1 pascal = 94 dB SPL.

The following list gives typical sensitivity specs for the three microphone types, in dBV/microbar:

Condenser: −65 dB (high sensitivity)

Moving coil: −75 dB (medium sensitivity)

Ribbon or small moving coil: −85 dB (low sensitivity)

Differences of a few dB among microphones are not critical.

The sensitivity of a microphone doesn't affect its sound quality. Rather, sensitivity affects the audibility of mixing-console noise (hiss). To achieve the same recording level, a low-sensitivity mic requires more mixer gain than a high-sensitivity mic, and more gain usually results in more noise. If you record quiet, distant instruments such as a classical guitar or chamber music, you'll hear more mixer noise with a low-sensitivity mic than with a high-sensitivity mic, all else being equal. Since stereo miking is usually done at a distance, high sensitivity is an asset.

Sensitivity is sometimes called "output level," but the two terms are not synonymous. Sensitivity is the output level produced by a particular input sound pressure level (SPL). The higher the SPL, the higher the output of any microphone.

SELF-NOISE

Self-noise is the electrical noise (hiss) a microphone produces. The microphone is put in a soundproof box, and its output noise voltage is measured. Self-noise is specified as the dB SPL of a source that would produce the same output voltage as the noise.

The self-noise spec is usually A-weighted; that is, the noise was measured through a filter that makes the measurement correlate more closely with the annoyance value. The filter rolls off low and high frequencies to simulate the frequency response of the human ear. An A-weighted self-noise spec of 20 dB SPL or less is excellent (quiet); a spec around 30 dB SPL is good, and a spec around 40 dB SPL is fair. A very quiet condenser mic has a self-noise spec of around 14 dBA (14 dB SPL, A-weighted).

SIGNAL-TO-NOISE RATIO (S/N)

Although referred to as a ratio, this is the difference between SPL and self-noise, all in dB. The higher the SPL of the sound source at the mic, or the lower the self-noise of the mic, the higher the S/N "ratio." For example, if the sound pressure level is 94 dB at the microphone, and the mic self-noise is 24 dB, the S/N ratio is 70 dB.

The higher the S/N ratio, the cleaner (more noise-free) is the signal. Given an SPL of 94 dB, a signal-to-noise ratio of 74 dB is excellent; 64 dB is good.

MICROPHONE TYPES

Another specification is the type of microphone: the generic classification. Some types of microphones for stereo miking are free-field, boundary, stereo, shotgun, and parabolic.

☐ Free-Field Microphone

Most microphones are of this type. They are meant to be used in a free-field; that is, away from reflective surfaces.

☐ Boundary Microphone

A boundary microphone is designed to be used on surfaces such as a floor, wall, table, piano lid, baffle, or panel. One example of a boundary microphone is the Crown Pressure Zone Microphone (PZM)® (shown in Figure 1-6). It includes a miniature electret condenser capsule mounted face down next to a sound-reflecting plate or boundary. Because of this construction, the microphone diaphragm receives direct and reflected

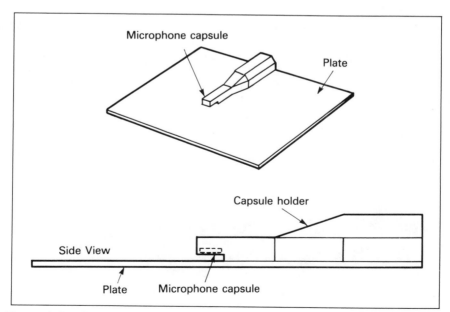

Figure 1-6 Crown PZM® construction. (Courtesy of Crown International.)

sounds in-phase at all frequencies, avoiding phase interference between them. The claimed benefits are a wide, smooth frequency response free of phase cancellations, excellent clarity and "reach," a hemispherical polar pattern, and uniform frequency response anywhere around the microphone. Because of this last characteristic, hall reverberation is picked up without tonal coloration.

If an omnidirectional boundary mic is placed on a panel, it becomes directional. Thus, boundary mics on angled panels can be used for stereo arrays. Boundary microphones are also available with a unidirectional polar pattern. They have the benefits of both boundary mounting and the unidirectional pattern. Such microphones are well suited for stage-floor pickup of drama, musicals, or small musical ensembles.

☐ Stereo Microphone

A stereo microphone combines two mic capsules in a single housing for convenient stereo recording. Simply place the microphone about 10 to 15 ft in front of a band, choir, or orchestra, and you'll get a stereo recording with little fuss. Several models of stereo microphones are listed in Chapter 9. One is shown in Figure 1-7.

Most stereo microphones are made with coincident microphone capsules. Since there is no horizontal spacing between the capsules, there also is no delay or phase shift between their signals. Thus the microphone is mono-compatible: the frequency response is the same in mono as in stereo. That's because there are no phase cancellations if the two channels are combined.

Stereo microphones are available in many configurations, such as XY, MS, Blumlein, ORTF, OSS, Soundfield, and SASS (these types are described in Chapters 4 and 5). Unlike the other types, the MS stereo microphone and Soundfield microphone let you remote-control the stereo spread and vary the stereo spread after recording. In general, a stereo microphone is easier to set up than two separate microphones, but it is more expensive.

Figure 1-7 Stereo microphone, a Neumann USM 69. (Courtesy of Gotham Audio Corporation.)

☐ Shotgun Microphone

A shotgun or line microphone is a long, tube-shaped microphone with a highly directional pickup pattern. It is used for highly selective pickup, or maximum rejection of background noise. Typical uses are for film/video dialog, news gathering, and outdoor recording. A stereo shotgun microphone is a mid-side type with a shotgun for the mid element and a bidirectional for the side element.

☐ Parabolic Microphone

This type of microphone has a large parabolic dish, or reflector, to focus sound on the microphone element. The parabolic type is even more directional than the shotgun type, but has a rougher, narrower frequency response. Microphones with big parabolic reflectors are directional down to lower frequencies than microphones with small reflectors. Also, the bigger the reflector, the lower the frequency response extends. There are no stereo parabolic microphones commercially available, but environmental recordist Dan Gibson has constructed his own to make some excellent recordings.

MICROPHONE ACCESSORIES

Various accessories used with microphones enhance their convenience, aid in placement, or reduce vibration pickup.

☐ Stands and Booms

These adjustable devices hold the microphones and let you position them as desired. A microphone stand has a heavy metal base that supports a vertical pipe. At the top of the pipe is a rotating clutch that lets you adjust the height of a smaller telescoping pipe inside the larger one. The top of the small pipe has a standard $\frac{5}{8}$"-27 thread, which screws into a microphone stand adapter. Camera stores have photographic stands, which are collapsible and lightweight—ideal for recording on-location. The thread is usually $\frac{1}{4}$"-20, which requires an adapter to fit a $\frac{5}{8}$"-27 thread in a mic stand adapter.

A boom is a long pipe that attaches to a mic stand. The angle and length of the boom are adjustable. The end of the boom is threaded to accept a microphone stand adapter, and the opposite end is weighted to balance the weight of the microphone. You can use a boom to raise a microphone farther off the floor, in order to stereo-mike an orchestra, for example.

☐ Stereo Microphone Adapter

A stereo microphone adapter, stereo bar, or stereo rail mounts two microphones on a single stand for convenient coincident and near-coincident stereo miking. Several models of these are listed in Chapter 9, and one is shown in Figure 1-8. In most models, the microphone spacing and angling are adjustable.

☐ Shock Mount

This device mounts on a microphone stand and holds a microphone in a resilient suspension to isolate the microphone from mechanical vibrations such as stand and floor thumps. The shock mount acts as a spring that resonates at a sub-audible frequency with the mass of the microphone. This mass-spring system attenuates mechanical vibrations above its resonance frequency.

Many microphones have an internal shock mount that isolates the microphone capsule from its housing; this reduces handling noise as well as stand thumps.

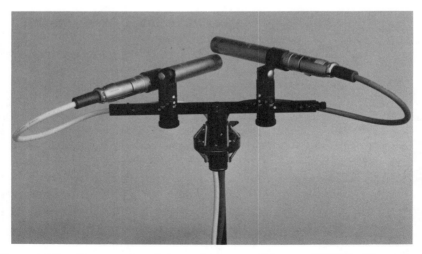

Figure 1-8 Stereo microphone adapter, a Schoeps UMS20. (Courtesy of Posthorn Recordings.)

☐ Phantom-Power Supply

Condenser microphones need power to operate their internal circuitry. Some use a battery; others use a remote phantom-power supply. This supply can be a stand-alone box, which you connect between the mic and your mixer mic input. Some mixing consoles have phantom power built in, available at each mic connector. Phantom power is usually 48 V DC on pins 2 and 3, with respect to pin 1.

Phantom power is supplied to the mic through its two-conductor shielded audio cable. The microphone receives power from and sends audio to the mixer along the same cable conductors.

☐ Junction Box and Snake

If you're recording with more than three microphones, you might want to plug them into a junction box with multiple connectors. A single thick multiconductor cable, called a *snake*, carries the signals from the junction box to your mixer. At the mixer end, the cable divides into several mic connectors that plug into the mixer.

☐ Splitter

You might need to send your microphone signals simultaneously to a broadcast mixer, recording mixer, and sound-reinforcement mixer. A microphone splitter does the job. It has one input for each microphone and two or three isolated outputs per microphone to feed each mixer. This device is passive; a distribution amplifier with the same function is active (it has amplification).

With a good grasp of microphone characteristics and accessories, we're now ready to discuss stereo miking theory.

REFERENCE

This chapter is based on B. Bartlett, "Microphones," in *Introduction to Professional Recording Techniques*, ed. by John Woram, Howard W. Sams & Co., Indianapolis, 1987, Chapter 6.

2

Overview of Stereo Microphone Techniques

Stereo microphone techniques are used mainly to record classical-music ensembles and soloists on location. These methods capture a sonic event as a whole, typically using only two or three microphones. During playback of a stereo recording, images of the musical instruments are heard in various locations between the stereo speakers. These images are in the same places, left-to-right, that the instruments were at the recording session. In addition, true-stereo miking conveys

- the depth or distance of each instrument
- the distance of the ensemble from the listener (the perspective)
- the spatial sense of the acoustic environment—the ambience or hall reverberation

WHY RECORD IN STEREO?

When planning a recording session, you may ask yourself, "Should I record in stereo with just a few mics? Or should I use several microphones placed close to the instruments and mix them with a mixer?"

Stereo miking is preferred for classical music, such as a symphony performed in a concert hall or a string quartet piece played in a recital hall. For classical-music recording, stereo methods have several advantages over close-mic methods.

For example, I said that stereo miking preserves depth, perspective, and hall ambience—all part of the sound of classical music as heard by the audience. These characteristics are lost with multiple closeup pan-potted microphones. But with a good stereo recording, you get a sense of an ensemble of musicians playing together in a shared space. Also, a pair of mics at a distance relays instrument timbres more accurately than closeup mics. Close-miked instruments in a classical setting sound too bright, edgy, or detailed compared to how they sound in the audience area.

Another advantage of stereo miking is that it tends to preserve the ensemble balance as intended by the composer. The composer has assigned dynamics (loudness notations) to the various instruments in order to produce a pleasing ensemble balance in the audience area. Thus, the correct balance or mix of the ensemble occurs at a distance, where all the instruments blend together acoustically. But this balance can be upset with multiple miking; you must rely on your own judgment (and the conductor's) regarding mixer settings to produce the composer's intended balance. Of course, even a stereo pair of mics can yield a faulty balance. But a stereo pair, being at a distance, is more likely to reproduce the balance as the audience hears it.

OTHER APPLICATIONS FOR STEREO MIKING

In contrast to a classical-music recording, a pop-music recording is made with multiple close mics because it sounds tighter and cleaner, which is the preferred style of production for pop music. Close miking also lets you experiment with multitrack mixes after the session. Still, stereo miking can be used in pop-music sessions for large sound sources within the ensemble, such as

- groups of singers
- piano
- drum-set cymbals overhead
- vibraphone, xylophone, and other percussion instruments
- string and horn sections

Other uses for stereo miking are

- samples
- sound effects
- background ambience and stereo dialog for film, video, and electronic news gathering
- audience reaction
- sports broadcasts
- radio group discussions
- radio plays
- drama

GOALS OF STEREO MIKING

Let's focus now on stereo-miking a large musical ensemble and define what we want to achieve. One objective is accurate localization. That is, the reproduced instruments should appear in the same relative locations as they were in the live performance. When this is achieved, instruments in the center of the ensemble are accurately reproduced midway between the two playback speakers. Instruments at the sides of the ensemble are reproduced from the left or right speaker. Instruments located half-way to one side are reproduced half-way to one side, and so on.

Figure 2-1 shows three stereo localization effects. In Figure 2-1(a), various instrument positions in an orchestra are shown: left, left-center, center, right-center, right. In Figure 2-1(b), the reproduced images of these instruments are accurately localized between the stereo pair of speakers. The stereo spread or stage width extends from speaker to speaker. If the microphones are placed improperly, the effect is either the narrow stage width shown in Figure 2-1(c) or the exaggerated separation shown in Figure 2-1(d). (Note that a large ensemble should spread from speaker to speaker, while a quartet can have a narrower spread.)

To judge these stereo localization effects, it's important to position yourself properly with respect to the monitor speakers. Sit as far from

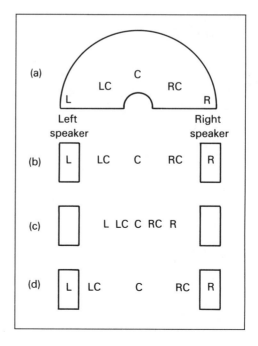

Figure 2-1 Stereo localization effects. (a) Orchestra instrument locations (top view). (b) Images accurately localized between speakers (the listener's perception). (c) Narrow stage-width effect. (d) Exaggerated separation effect.

Overview of Stereo Microphone Techniques □ 17

the speakers as they are spaced apart. The speakers will appear to be 60° apart, which is about the same angle an orchestra fills when viewed from the typical ideal seat in the audience (say, tenth row center). Sit exactly between the speakers (equidistant from them); otherwise, the images will shift toward the side on which you're sitting and will become less sharp. Also, pull out the speakers several feet from the walls: this delays and weakens early reflections which can degrade stereo imaging.

The reproduced size of an instrument or instrumental section should match its size in real life. A guitar should be a point source; a piano or string section should have some stereo spread. Each instrument's location should be as clearly defined as it was in the concert hall, as heard from the ideal seat. Some argue that the reproduced images should be sharper than in real life to supplant the missing visual cues—in other words, since you can't see the instruments during loudspeaker reproduction, extra-sharp images might enhance the realism.

The reproduced reverberation (concert-hall ambience) should either surround the listener, or at least it should spread evenly between the speakers (as shown in Figure 2-2). Typical stereo miking techniques

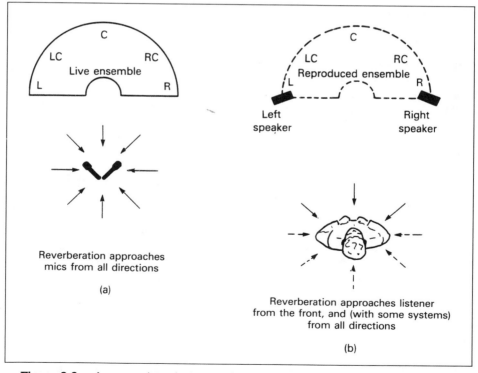

Figure 2-2 Accurate imaging: sound source location and size are reproduced during playback, as well as the reverberant field. (a) Recording. (b) Playback. (Courtesy of Howard Sams, Inc.)

reproduce the hall reverberation up front, in a line between the speakers, so you don't get a sense of being immersed in the hall ambience. To make the recorded reverberation surround the listener, you need extra speakers to the side or rear, an add-on reverberation simulator, or a head-related crosstalk canceller (explained in Chapter 6). However, spaced-microphone recordings artificially produce a sense of some surrounding reverberation.

There should also be a sense of stage depth. Front-row instruments should sound closer than back-row instruments.

■
TYPES OF STEREO MICROPHONE TECHNIQUES

There are three general microphone techniques commonly used for stereo recording: the coincident-pair, the spaced-pair, and the near-coincident-pair techniques [1–3]. A fourth technique uses baffled omni's or an artificial head, covered in Chapter 6. Let's look at the first three techniques in detail.

☐ Coincident Pair

With the coincident-pair method (XY or intensity stereo method), two directional microphones are mounted with their grilles nearly touching and their diaphragms placed one above the other, angled apart to aim approximately toward the left and right sides of the ensemble (as in Figure 2-3). For example, two cardioid microphones can be mounted angled apart, their grilles one above the other. Other directional patterns can be used, too. The greater the angle between microphones, and the narrower the polar pattern, the wider the stereo spread.

Let me explain how the coincident-pair technique produces localizable images. As described in the first chapter, a directional microphone is most sensitive to sounds in front of the microphone (on-axis) and progressively less sensitive to sounds arriving off-axis. That is, a directional mic produces a relatively high-level signal from the sound source it's aimed at and a relatively low-level signal for all other sound sources.

The coincident-pair method uses two directional mics symmetrically angled from the center line, as in Figure 2-3. Instruments in the center of the ensemble produce an identical signal from each microphone. During playback, an image of the center instruments is heard midway between the stereo pair of loudspeakers. That's because identical signals in each channel produce a centrally located image.

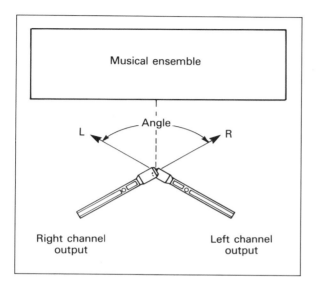

Figure 2-3
Coincident-pair technique.

If an instrument is off-center to the right, it is more on-axis to the right-aiming mic than to the left-aiming mic; thus, the right mic will produce a higher-level signal than the left mic. During playback of this recording, the right speaker will play at a higher level than the left speaker, reproducing the image off-center to the right—where the instrument was during recording.

The coincident array codes instrument positions into level differences (intensity or amplitude differences) between channels. During playback, the brain decodes these level differences back into corresponding image locations. A pan pot in a mixing console works on the same principle.

If one channel is 15 to 20 dB louder than the other, the image shifts all the way to the louder speaker. So, if we want the right side of the orchestra to be reproduced at the right speaker, the right side of the orchestra must produce a signal level 20 dB higher from the right mic than from the left mic. This occurs when the mics are angled apart sufficiently. The correct angle depends on the polar pattern. Instruments part-way off-center produce interchannel level differences less than 20 dB, so they are reproduced part-way off-center.

Listening tests have shown that coincident cardioid microphones tend to reproduce the musical ensemble with a narrow stereo spread. That is, the reproduced ensemble does not spread all the way between speakers.

A coincident-pair method with excellent localization is the Blumlein array, which uses two bidirectional mics angled 90° apart and facing the left and right sides of the ensemble.

A special form of the coincident-pair technique is the mid-side (MS) recording method illustrated in Figure 2-4. A microphone facing the

20 □ STEREO MICROPHONE TECHNIQUES

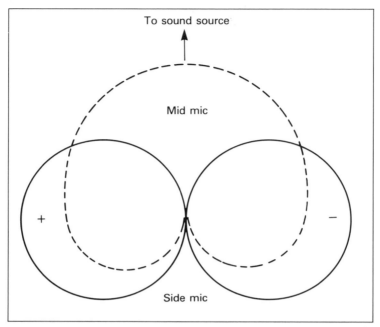

Figure 2-4 Mid-side technique. Left channel = mid + side. Right channel = mid − side. The polarity of the side mic lobes is indicated by + and −.

middle of the orchestra is summed and differenced with a bidirectional microphone aiming to the sides. This produces left- and right-channel signals. With this technique, the stereo spread can be remote-controlled by varying the ratio of the mid signal to the side signal. This remote control is useful at live concerts, where you can't physically adjust the microphones during the concert. MS localization accuracy is excellent.

Mid-side recordings are sometimes said to lack spaciousness. But this can be improved with spatial equalization, in which a circuit boosts the bass 4 dB (+2 dB at 600 Hz) in the L − R, or side, signal, with a corresponding cut in the L + R, or mid, signal [4]. Another way to improve the spaciousness is to mix in a distant MS microphone—one set about 30 to 75 ft from the main MS microphone.

Two coincident microphone capsules can be mounted in a single housing for convenience: this forms a stereo microphone. Several of these are listed in Chapter 9.

A recording made with coincident techniques is mono-compatible, that is, the frequency response is the same in mono or stereo. Because of the coincident placement, there is no time or phase difference between channels to degrade the frequency response if both channels are combined to mono. If you expect your recordings to be heard in mono (say, on radio, TV, or film), you should consider coincident methods.

☐ Spaced Pair

With the spaced-pair (or A-B) technique, two identical microphones are placed several feet apart, aiming straight ahead toward the musical ensemble (as in Figure 2-5). The mics can have any polar pattern, but the omnidirectional pattern is the most popular for this method. The greater the spacing between the microphones, the greater the stereo spread.

Instruments in the center of the ensemble produce an identical signal from each microphone. During playback of this recording, an image of the center instruments is heard midway between the stereo pair of loudspeakers.

If an instrument is off-center, it is closer to one mic than the other, so its sound reaches the closer microphone before it reaches the other one. Consequently, the microphones produce an approximately identical signal, except that one mic signal is delayed with respect to the other. If you send an identical signal to two stereo speakers with one channel delayed, the sound image shifts off center. With a spaced-pair recording, off-center instruments produce a delay in one mic channel, so they are reproduced off-center.

The spaced-pair array codes instrument positions into time differences between channels. During playback, the brain decodes these time differences back into corresponding image locations. It takes only about 1.5 milliseconds (msec) of delay to shift an image all the way to one

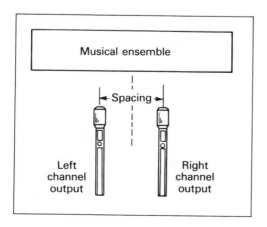

Figure 2-5 Spaced-pair technique.

speaker, so if we want the right side of the orchestra to be reproduced at the right speaker, its sound must arrive at the right mic about 1.5 msec before it reaches the left mic. In other words, the mics should be spaced about 2 ft apart, because this spacing produces the appropriate delay to place right-side instruments at the right speaker. Instruments part-way off-center produce interchannel delays less than 1.5 msec, so they are reproduced part-way off-center.

If the spacing between mics is, say, 12 ft, instruments slightly off-center produce interchannel delays greater than 1.5 msec, which places their images at the left or right speaker. This is called an "exaggerated separation" or "ping pong" effect.

On the other hand, if the mics are too close together, the delays produced will be inadequate to provide much stereo spread. In addition, the mics will tend to favor the center of the ensemble because the mics are closest to the center instruments.

To record a good musical balance, we need to place the mics about 10 or 12 ft apart, but such a spacing results in exaggerated separation. One solution is to place a third microphone midway between the original pair and mix its output to both channels. That way, the ensemble is recorded with a good balance, and the stereo spread is not exaggerated.

The spaced-pair method tends to make off-center images relatively unfocused or hard to localize, for this reason: Spaced-microphone recordings have time differences between channels, and stereo images produced solely by time differences are relatively unfocused. Centered instruments are still heard clearly in the center, but off-center instruments are difficult to pinpoint between speakers. This method is useful if you prefer the sonic images to be diffuse, rather than sharply focused (say, for a blended effect).

There's another problem with spaced microphones. The large time differences between channels correspond to gross phase differences between channels. Out-of-phase low-frequency signals can cause excessive vertical modulation of the record groove, making records difficult to cut unless the cutting level or low-frequency stereo separation is reduced. In addition, combining both mics to mono sometimes causes phase cancellations of various frequencies, which may or may not be audible.

There is an advantage with spaced miking, however. Spaced microphones are said to provide a "warm" sense of ambience, in which concert hall reverberation seems to surround the instruments and, sometimes, the listener. Here's why: the two channels of recorded reverberant sound are incoherent; that is, they have random phase relationships. Incoherent signals from stereo loudspeakers sound diffuse and spacious. Since reverberation is picked up and reproduced incoherently by spaced microphones, it sounds diffuse and spacious. The simulated spaciousness caused by the phasiness is not necessarily realistic [5], but it is pleasant to many listeners.

Overview of Stereo Microphone Techniques □ 23

Another advantage of the spaced-microphone technique is the ability to use omnidirectional microphones. An omnidirectional condenser microphone has more extended low-frequency response than a unidirectional condenser microphone and tends to have a smoother response and less off-axis coloration. (Microphone characteristics are explained in detail in the first chapter.)

□ Near-Coincident Pair

As shown in Figure 2-6, the near-coincident technique uses two directional microphones angled apart, with their grilles horizontally spaced a few inches apart. Even a few inches of spacing increases the stereo spread and adds a sense of depth and airiness to the recording. The greater the angle or spacing between mics, the greater the stereo spread.

Here's how this method works: angling directional mics produces level differences between channels; spacing mics produces time differences. The interchannel level differences and time differences combine to create the stereo effect. If the angling or spacing is too great, the result is exaggerated separation. If the angling or spacing is too small, the result is a narrow stereo spread.

The most common example of the near-coincident method is the ORTF system, which uses two cardioids angled 110° apart and spaced seven

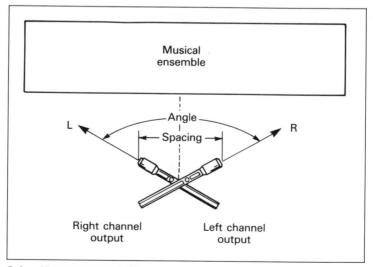

Figure 2-6 Near-coincident-pair technique.

inches (17 cm) apart horizontally. (ORTF stands for Office de Radiodiffusion Television Française—French Broadcasting Organization.) This method tends to provide accurate localization; that is, instruments at the sides of the orchestra are reproduced at or very near the speakers, and instruments half-way to one side tend to be reproduced half-way to one side.

COMPARING THE THREE STEREO MIKING TECHNIQUES

The coincident-pair technique has the following features:

- It uses two directional mics angled apart with grilles nearly touching, one mic's diaphragm above the other.
- Level differences between channels produce the stereo effect.
- Images are sharp.
- Stereo spread ranges from narrow to accurate.
- Signals are mono-compatible.

The spaced-pair technique has these features:

- It uses two mics spaced several feet apart.
- Time differences between channels produce the stereo effect.
- Off-center images are diffuse.
- Stereo spread tends to be exaggerated unless a third center mic is used.
- It provides a warm sense of ambience.
- It may cause record-cutting problems.

The near-coincident-pair technique has these features:

- It uses two directional mics angled apart and spaced a few inches apart.
- Level and time differences between channels produce the stereo effect.
- Images are sharp.
- Stereo spread tends to be accurate.
- It provides a greater sense of "air" and depth than coincident methods.

MOUNTING HARDWARE

With coincident and near-coincident techniques, the microphones should be rigidly mounted with respect to one another, so that they can be moved as a unit without disturbing their arrangement. A device for this purpose is called a stereo microphone adapter or stereo bar. It mounts two microphones on a single stand, and microphone angling and spacing are adjustable. A number of these devices are listed in Chapter 9.

MICROPHONE REQUIREMENTS

The sound source dictates the requirements of the recording microphones. Most acoustic instruments produce frequencies from about 40 Hz (string bass and bass drum) to about 20,000 Hz (cymbals, castanets, triangles). A microphone with uniform response between these frequency limits will do full justice to the music.

The highest octave from 10 kHz to 20 kHz adds transparency, air, and realism to the recording. You may need to roll off (filter out) frequencies below 80 Hz to eliminate rumble from trucks and air conditioning, unless you want to record organ or bass-drum fundamentals.

Sound from an orchestra or band approaches each microphone from a broad range of angles. To reproduce all the instruments' timbres equally well, the microphone should have a broad, flat response at all angles of incidence within at least ±90°, that is, the polar pattern should be uniform with frequency. Microphones with small-diameter diaphragms usually meet this requirement best. (Note that some microphones have small diaphragms inside large housings.)

If you're forced to record at a great distance, a frequency response elevated up to 4 dB above 4 kHz might sound more natural than a flat response. Another benefit of a rising high end is that you can roll it off in post production, reducing analog tape hiss. Since classical music covers a wide dynamic range (up to 80 dB), the recording microphones should have very low noise and distortion. In distant-miking applications, the sensitivity should be high to override mixer noise.

For sharp imaging, the microphone pair should be well matched in frequency response, phase response, and polar pattern.

We've investigated several microphone arrangements for recording in stereo. Each has its advantages and disadvantages. Which method you choose depends on the sonic compromises you are willing to make.

REFERENCES

1. B. Bartlett, "Stereo Microphone Technique," *db*, Vol. 13, No. 12, 1979 (Dec.), pp. 34–46.
2. J. Eargle, "Stereo Microphone Techniques," in *The Microphone Handbook*, Elar Publishing, Plainview, NY, 1981, Chapter 10.
3. R. Streicher and W. Dooley, "Basic Stereo Microphone Perspectives—A Review," *Journal of the Audio Engineering Society*, Vol. 33, No. 7/8, 1984 (July/Aug.), pp. 548–556.
4. David Griesinger, "Spaciousness and Localization in Listening Rooms and Their Effects on the Recording Technique," *J. Audio Eng. Soc.*, Vol. 34, No. 4, 1986 (April), pp. 255–268.
5. S. Lipshitz, "Stereo Microphone Techniques: Are the Purists Wrong?" *J. Audio Eng. Soc.*, Vol. 34, No. 9, 1986 (September), pp. 716–744.

3

Stereo Imaging Theory

☐

A sound system with good stereo imaging can form apparent sources of sound, such as reproduced musical instruments, in well-defined locations—usually between a pair of loudspeakers placed in front of the listener. These apparent sound sources are called *images*.

This chapter explains

- terms related to stereo imaging
- how we localize real sound sources
- how we localize images
- how microphone placement controls image location

You can use stereo microphone techniques without reading this chapter. However, if you want to deepen your understanding of what's going on or develop your own stereo array, it's worthwhile to study the theory and simple math in this chapter.

DEFINITIONS

First, I'll define several terms related to stereo imaging.

Fusion refers to the synthesis of a single apparent source of sound (an image or "phantom image") from two or more real sound sources (such as loudspeakers).

The *location* of an image is its angular position relative to a point straight ahead of a listener, or its position relative to the loudspeakers. This is shown in Figure 3-1. An aim of high fidelity is to reproduce the images in the locations intended by the recording engineer or producer. In some productions—usually classical-music recordings—a goal of the recording engineer or producer is to place the images in the same relative locations as the instruments were during the live performance.

Stereo spread or *stage width* (Figure 3-2) is the distance between the extreme left and right images of a reproduced ensemble of instru-

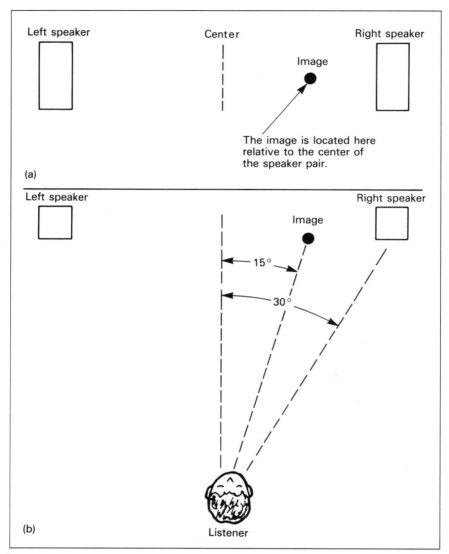

Figure 3-1 Example of image location: (a) listener's view; (b) top view.

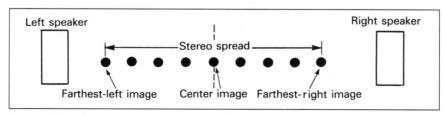

Figure 3-2 Stereo spread or stage width.

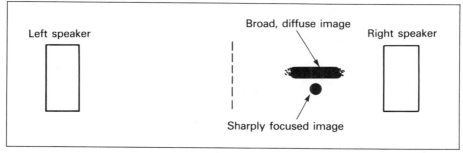

Figure 3-3 Image focus or size (listener's perception).

ments. The stereo spread is wide if the ensemble appears to spread all the way between a pair of loudspeakers. The spread is narrow if the reproduced ensemble occupies only a small space between the speakers. Sometimes the reproduced reverberation or ambience spreads from speaker to speaker even when the reproduced ensemble width is narrow.

Image focus or *size* (Figure 3-3) refers to the degree of fusion of an image, or its positional definition. A sharply focused image is described as being pinpointed, precise, narrow, sharp, resolved, well-defined, or easy to localize. A poorly focused image is hard to localize, spread, broad, smeared, vague, diffuse. A *natural image* is focused to the same degree as the real instrument being reproduced.

Depth is the apparent distance of an image from a listener, a sense of closeness and farness of various instruments.

Elevation is an image displacement in height above the speaker plane.

Image movement is a reproduction of the movement of the sound source, if any. The image should not move unaccountably.

Localization is the ability of a listener to tell the direction of a sound. It is also the relation between interchannel or interaural differences and perceived image location.

HOW WE LOCALIZE REAL SOUND SOURCES

The human hearing system uses the direct sound and early reflections to localize a sound source. The direct sound and reflections within about 2 msec contribute to localization [1, 2]. Reflections occurring up to 5 to 35 msec after the direct sound influence image broadening [3]. Echoes delayed more than about 5 to 50 msec (depending on program material)

30 □ STEREO MICROPHONE TECHNIQUES

do not fuse in time with the early sound, but do contribute to perceived tonal spectrum [4].

Imagine a sound source and a listener. Let's say that the source is in front of the listener and to the left of center (as in Figure 3-4). Sound travels a longer distance to the right ear than to the left ear, so the sound arrives at the right ear after it arrives at the left ear. In other words, the right-ear signal is delayed relative to the left-ear signal. Every source location produces a unique arrival-time difference between ears [5].

In addition, the head acts as an obstacle to sounds above about 1000 Hz. High frequencies are shadowed by the head, so a different spectrum (amplitude versus frequency) appears at each ear [6, 7]. Every source location produces a unique spectral difference between ears (Figure 3-5).

We have learned to associate certain interaural differences with specific directions of the sound source. When presented with a new source location, we match what we hear with a memorized pattern of a similar situation to determine direction [8].

As stated before, an important localization cue is the interaural arrival-time difference of the signal envelope. We perceive this difference at any *change* in the sound—a transient, a pause, or a change in timbre. For this reason, we localize transients more easily than continuous sounds [9]. The time difference between ear signals can also be considered as a phase difference between sound waves arriving at the ears (Figure 3-6). This phase shift rises with frequency.

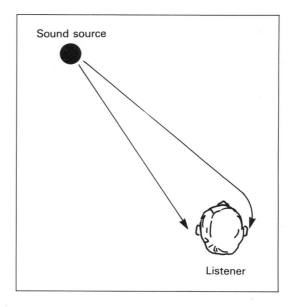

Figure 3-4 Sound traveling from a source to a listener's ears.

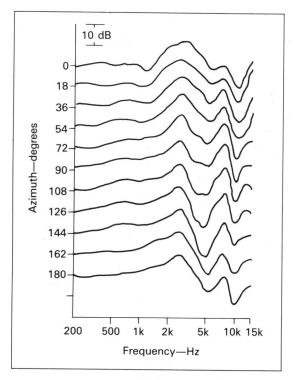

Figure 3-5 Frequency response of the ear at different azimuth angles. 0° is straight ahead; 90° is to the side of the ear being measured; 180° is behind the head (after reference [7]).

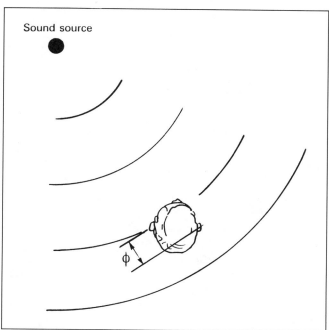

Figure 3-6 Phase shift ϕ between sound waves at the ears.

When sound waves from a real source strike a listener's head, a different spectrum of amplitude and phase appears at each ear. These interaural differences are translated by the brain into a perceived direction of the sound source. Every direction is associated with a different set of interaural differences.

The ears make use of interaural phase differences to localize sounds between about 100 Hz and 700 Hz. Frequencies below about 100 Hz are not localized (making "subwoofer/satellite" speaker systems feasible) [10]. Above about 1500 Hz, amplitude differences between ears contribute to localization. Between about 700 and 1500 Hz, both phase and amplitude differences are used to tell the direction of a sound [11, 12].

Small movements of the head change the arrival-time difference at the ears. The brain uses this information as another cue for source location [13], especially for distance and front/back discrimination.

The outer ears (pinnae) play a part as well [14]. In each pinna, sound reflections from various ridges combine with the direct sound, causing phase cancellations—frequency notches in the perceived spectrum. The positions of the notches in the spectrum vary with the source height. We perceive these notch patterns not as a tonal coloration, but as height information. Also, we can discriminate sounds in front from those in back because of the pinnae's shadowing effect at high frequencies.

Some of the cues used by the ears can be omitted without destroying localization accuracy if other cues are still present.

HOW WE LOCALIZE IMAGES BETWEEN SPEAKERS

Now that we've discussed how we localize real sound sources, let's look at how we localize their images reproduced over loudspeakers. Imagine that you're sitting between two stereo speakers as in Figure 3-7. If you feed a musical signal equally to both channels in the same polarity, you'll perceive an image between the two speakers. Normally you'll hear a single synthetic source, rather than two separate loudspeaker sources.

Each ear hears both speakers. For example, the left ear hears the left speaker signal, then, after a short delay due to the longer travel path, hears the right speaker signal. At each ear, the signals from both speakers sum or add together vectorially to produce a resultant signal.

Suppose that we make the signal louder in one speaker. That is, we create a level difference between the speakers. Surprisingly, this causes an arrival-time difference at the ears [15]. This is a result of the phasor addition of both speaker signals at each ear.

Stereo Imaging Theory □ 33

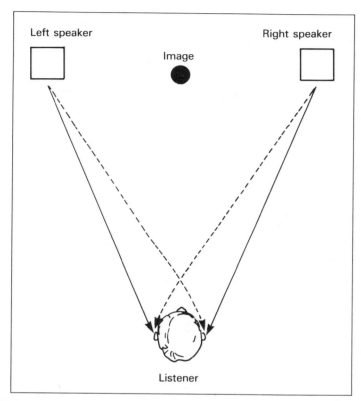

Figure 3-7   Two ears receiving signals from two speakers.

Remember to distinguish *interchannel* differences (between speaker channels) from *interaural* differences (between ears). An interchannel *level* difference does not appear as an interaural level difference, but rather as an interaural *time* difference.

We can use this speaker-generated interaural time difference to place images. Here's how: suppose we want to place an image 15° to one side. A real sound source 15° to one side produces an interaural time difference of 0.13 msec. If we can make the speakers produce an interaural time difference of 0.13 msec, we'll hear the image 15° to one side. We can fool the hearing system into believing there's a real source at that angle. This occurs when the speakers differ in level by a certain amount.

The polarity of the two channels affects localization as well. To explain polarity: if the signals sent to two speaker channels are in polarity, they are in phase at all frequencies; they both go positive in voltage at the same time. If the signals are out of polarity, they are 180° out of phase at all frequencies. One channel's signal goes positive when the other channel's signal goes negative. Opposite-polarity signals are sometimes incorrectly referred to as being "out of phase."

If the signals are in opposite polarity between channels and equal level in both channels, the resulting image has a diffuse, directionless quality and cannot be localized. If the signals are in opposite polarity, and higher level in one channel than the other, the image often appears outside the bounds of the speaker pair. You'd hear an image left of the left speaker or right of the right speaker [16].

Opposite polarity can occur in several ways. Two microphones are opposite polarity if the wires to connector pins 2 and 3 are reversed in one microphone. Two speakers are opposite polarity if the speaker-cable leads are reversed at one speaker. A single microphone might have different parts of its polar pattern in opposite polarity. For example, the rear lobe of a bidirectional pattern is opposite in polarity to the front lobe. If sound from a particular direction reaches the front lobe of the left-channel mic and the rear lobe of the right-channel mic, the two channels will be opposite polarity. The resulting image of that sound source will either be diffuse, or outside the speaker pair.

REQUIREMENTS FOR NATURAL IMAGING OVER LOUDSPEAKERS

To the extent that a sound recording-and-reproducing system can duplicate the interaural differences produced by a real source, the resultant image will be accurately localized. In other words, when reproduced sounds reaching a listener's ears have amplitude and phase differences corresponding to those of a real sound source at some particular angle, the listener perceives a well-fused, naturally focused image at that same angle. Conversely, when unnatural amplitude and phase relations are produced, the image appears relatively diffuse rather than sharp and is harder to localize accurately [17].

The required *interaural* differences for realistic imaging can be produced by certain *interchannel* differences. Placing an image in a precise location requires a particular amplitude difference versus frequency and phase difference versus frequency between channels. These have been calculated by Cooper and Bauck for several image angles [18]. Gerzon, Nakabayashi, and Koshigoe have calculated the interaural or interchannel differences required to produce any image direction at a single frequency [19–21].

Figure 3-8, for example, shows the interchannel differences required to place an image at 15° to the left of center when the speakers are placed ±30° in front of the listener [22].

As Figure 3-8 shows, the interchannel differences required for natural imaging vary with frequency. Specifically, Cooper and Bauck indicate that interchannel amplitude differences are needed below approximately 1700 Hz and interchannel time differences are needed above that frequency [23]. Specifically,

- At low frequencies, the amplitude difference needed for a 15° image angle is about 10 dB.
- Between 1.7 kHz and 5 kHz, the amplitude difference goes to approximately 0 dB.
- Above 1.7 kHz, the phase difference corresponds to a group delay (interchannel time difference) of about 0.547 msec, or 7.39 inches for a hypothetical spacing between microphones used for stereo recording.

This theory is based on the "shadowing" of sound traveling around a sphere. The description given here simplifies the complex requirements, but it conveys the basic idea. Cooper notes that "moderate deviations from these specifications might not lead to noticeable auditory distress or faulty imaging."

The Cooper-Bauck criteria can be met by recording with a dummy head whose signals are specially processed. A dummy head used for binaural recording is a modeled head with a flush-mounted microphone in each ear. Time and spectral differences between channels create the stereo images. (Spectral differences are amplitude differences that vary with frequency).

Although a dummy-head binaural recording can provide excellent imaging over headphones, it produces poor localization over loudspeakers at low frequencies [24] unless spatial equalization (a shuffler circuit) is used [25]. Spatial equalization boosts the low frequencies in the difference (L-R) signal.

Binaural recording can produce images surrounding a listener wearing headphones, but can produce only frontal images over loudspeakers, unless a transaural converter is used. A *transaural converter* is an electronic device that converts binaural signals (for headphone playback) into stereo signals (for loudspeaker playback). Transaural stereo is a method of surround-sound reproduction using a dummy head for binaural recording, processed electronically to remove head-related crosstalk when the recording is heard over two loudspeakers [26–35]. (More on this in Chapter 6.)

Cooper recommends that, for natural imaging, the speakers' interchannel differences be controlled so that their signals sum at the ears to produce the correct interaural differences. According to Theile [36],

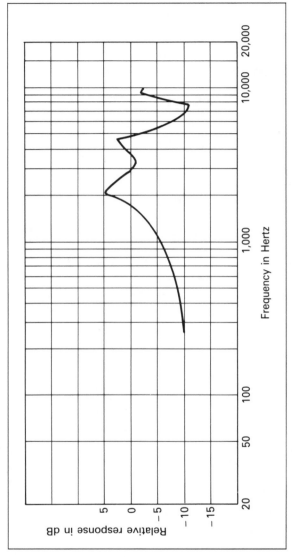

Figure 3.8a Amplitude of right channel relative to left channel, for image location 15° to the left of center when speakers are ±30° in front of listener.

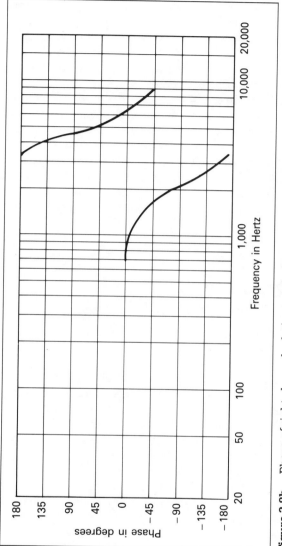

Figure 3.8b Phase of right channel relative to left channel, for image location 15° to the left of center.

Cooper's theory (based on summing localization) is in error because it applies only to sine waves and may not apply to broadband spectral effects. He proposes a different theory of localization, the association model. This theory suggests that, when listening to two stereo loudspeakers, we ignore our interaural differences and instead use the speakers' interchannel differences to localize images.

The interchannel differences needed for best stereo, Theile says, are head-related. The ideal stereo-miking technique would use, perhaps, two ear-spaced microphones flush-mounted in a head-sized sphere and equalized for flat subjective response. This would produce interchannel spectral and time differences, which, Theile claims, are optimum for stereo. The interchannel differences—time differences at low frequencies, amplitude differences at high frequencies—are the opposite of Cooper's requirements for natural stereo imaging! Time will tell which theory is closer to the truth.

CURRENTLY USED IMAGE-LOCALIZATION MECHANISMS

The ear can be fooled into hearing reasonably sharp images between speakers by less sophisticated signal processing. Simple amplitude and/or time differences between channels, constant with frequency, can produce localizable images. Bartlett, Jordan, Dutton, Cabot, Williams, Blauert, and Rumsey give test results showing image location as a function of interchannel amplitude or time differences [37–43]. Bartlett's results are shown later in this chapter.

For example, given a speech signal, if the left channel is 7.5 dB louder than the right channel, an image will appear at approximately 15° to the left of center when the speakers are placed ±30° in front of the listener. A delay in the right channel of about 0.5 msec will accomplish the same thing, although image locations produced solely by time differences are relatively vague and hard to localize.

Griesinger notes that pure interchannel time differences do not localize lowpass-filtered male speech below 500 Hz over loudspeakers. Amplitude (level) differences are needed to localize low-frequency sounds. Either amplitude or time differences can localize high-frequency sounds [44]. However, Blauert gives evidence by Wendt that interchannel time differences do cause localization at 327 Hz [45].

The interchannel differences produced in current stereo recordings are just simple approximations of what is required. Current practice makes use of interchannel amplitude differences and/or time differences to locate images. These differences are constant with frequency. Still, reasonably sharp images are produced. Let's look at exactly how these differences localize images [46].

☐ Localization by Amplitude Differences

The location of images between two loudspeakers depends in part on the signal amplitude differences between the loudspeakers. Suppose a speech signal is sent to two stereo loudspeakers, with the signal to each speaker identical except for an amplitude (level) difference (as shown in Figure 3-9). We'll create an amplitude difference by inserting an attenuator in one channel.

Figure 3-10 shows the approximate sound image location between speakers versus the amplitude difference between channels, in decibels. A 0 dB difference (equal level from each speaker) makes the image of the sound source appear in the center, midway between the speakers. Increasing the difference places the image farther away from the center. A difference of 15 to 20 dB makes the image appear at only one speaker.

(The information in this figure is based on carefully controlled listening tests. The data is the average of the responses of ten trained listeners. They auditioned a pair of signal-aligned, high-quality loud-

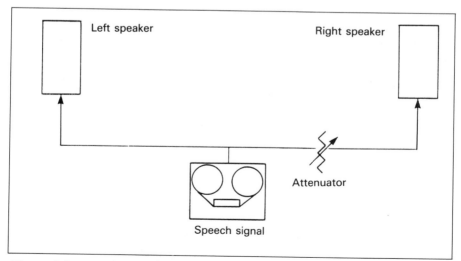

Figure 3-9 Sending a speech signal to two stereo loudspeakers with attenuation in one channel.

40 □ STEREO MICROPHONE TECHNIQUES

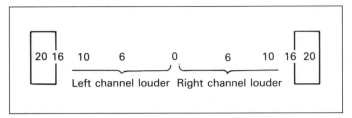

Figure 3-10 Stereo-image location versus amplitude difference between channels, in dB (listener's perception).

speakers several feet from the walls in the "typical" listening room, while sitting centered between the speakers at a 60° listening angle.)

How can we create this effect with a stereo microphone array? Suppose two cardioid microphones are crossed at 90° to each other, with the grille of one microphone directly above the other (Figure 3-11). The microphones are angled 45° to the left and right of the center of the orchestra. Sounds arriving from the center of the orchestra will be picked up equally by both microphones. During playback, there will be equal levels from both speakers and, consequently, a center image is produced.

Suppose that the extreme right side of the orchestra is 45° off-center, from the viewpoint of the microphone pair. Sounds arriving from the

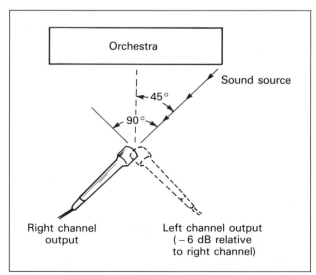

Figure 3-11 Cardioids crossed at 90°, picking up a source at one end of an orchestra.

extreme right side of the orchestra approach the right-aiming microphone on axis, but they will approach the left-aiming microphone at 90° off axis (as shown in Figure 3-11). A cardioid polar pattern has a 6 dB lower level at 90° off axis than it has on axis. So, the extreme-right sound source will produce a 6 dB lower output from the left microphone than from the right microphone.

Thus, we have a 6 dB amplitude difference between channels. According to Figure 3-10, the image of the extreme-right side of the orchestra will now be reproduced right-of-center. Instruments in between the center and the right side of the orchestra will be reproduced somewhere between the 0 dB point and the 6 dB point.

If we angle the microphones farther apart, for example 135°, the difference produced between channels for the same source is around 10 dB. As a result, the right-side stereo image will appear farther to the right than it did with 90° angling. (Note that it is not necessary to aim the microphones exactly at the left and right sides of the ensemble.)

The farther to one side a sound source is, the greater the amplitude difference between channels it produces, and so, the father from center is its reproduced sound image.

☐ Localization by Time Differences

Phantom-image location also depends on the signal time differences between channels. Suppose we send the same speech signal to two speakers at equal levels, but with one channel delayed (as in Figure 3-12).

Figure 3-13 shows the approximate sound image location between speakers, with various time differences between channels, in milliseconds. A 0 msec difference (no time difference between speaker channels) makes the image appear in the center. As the time difference increases, the phantom image appears farther off-center. A 1.5 msec difference or delay is sufficient to place the image at only one speaker.

Spacing two microphones apart horizontally—even by a few inches—produces a time difference between channels for off-center sources. A sound arriving from the right side of the orchestra will reach the right microphone first, simply because it is closer to the sound source (as in Figure 3-14). For example, if the sound source is 45° to the right, and the microphones are 8 inches apart, the time difference produced between channels for this source is about 0.4 msec. For the same source, a 20-inch spacing between microphones produces a 1.5 msec time difference between channels, placing the reproduced sound image at one speaker.

With spaced-pair microphones, the farther a sound source is from the center of the orchestra, the greater the time difference between channels and so, the farther from center is its reproduced sound image.

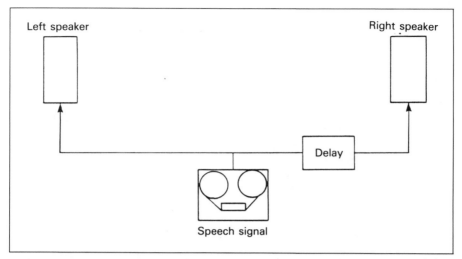

Figure 3-12 Sending a speech signal to two speakers with one channel delayed.

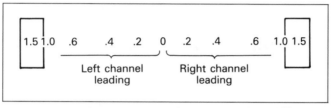

Figure 3-13 Approximate image location versus time difference between channels, in milliseconds (listener's perception).

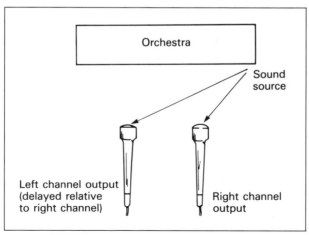

Figure 3-14 Microphones spaced apart, picking up a source at one end of an orchestra.

Stereo Imaging Theory □ 43

□ Localization by Amplitude and Time Differences

Phantom images can also be localized by a combination of amplitude and time differences. Suppose 90° angled cardioid microphones are spaced 8 inches apart (as in Figure 3-15). A sound source 45° to the right will produce a 6 dB level difference between channels, and a 0.4 msec difference between channels. The image shift of the 6 dB level difference adds to the image shift of the 0.4 msec difference to place the sound image at the right speaker. Certain other combinations of angling and spacing accomplish the same thing.

□ Summary

If a speech signal is recorded on two channels, its reproduced sound image will appear at only one speaker when

- the signal is 15 to 20 dB lower in one channel, or
- the signal is delayed 1.5 msec in one channel, or
- the signal in one channel is lower in level and delayed by a certain amount.

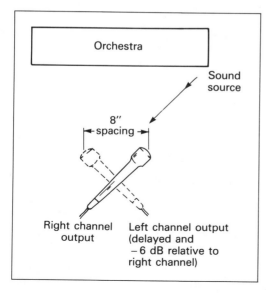

Figure 3-15 Cardioids angled 90° and spaced 8 inches, picking up a source at one end of an orchestra.

44 □ STEREO MICROPHONE TECHNIQUES

When amplitude and time differences are combined to place images, the sharpest imaging occurs when the channel that is lower in level is also the channel that is delayed. If the higher-level channel is delayed, image confusion results due to the conflicting time/amplitude cues.

We have seen that angling directional microphones (coincident placement) produces amplitude differences between channels. Spacing microphones (spaced-pair placement) produces time differences between channels. Angling and spacing directional microphones (near-coincident placement) produces both amplitude and time differences between channels. These differences localize the reproduced sound image between a pair of loudspeakers.

PREDICTING IMAGE LOCATIONS

Suppose you have a pair of microphones for stereo recording. Given their polar pattern, angling, and spacing, you can predict the interchannel amplitude and time differences for any source angle. Hence, you can predict the localization of any stereo microphone array.

This prediction assumes that the microphones have ideal polar patterns, and that these patterns do not vary with frequency. It's an unrealistic assumption, but the prediction agrees well with listening tests.

The amplitude difference between channels in dB is given by

$$\Delta dB = 20 \log \left[\frac{a + b \cos((\theta_m/2) - \theta_s)}{a + b \cos((\theta_m/2) + \theta_s)} \right] \quad (1)$$

where ΔdB = amplitude difference between channels, in dB
$a + b \cos(\theta)$ = polar equation for the microphone.

Omnidirectional:	$a = 1$	$b = 0$
Bidirectional:	$a = 0$	$b = 1$
Cardioid:	$a = 0.5$	$b = 0.5$
Supercardioid:	$a = 0.366$	$b = 0.634$
Hypercardioid:	$a = 0.25$	$b = 0.75$

θ_m = angle between microphone axes, in degrees
θ_s = source angle (how far off-center the sound source is, in degrees)

These variables are shown in Figure 3-16.

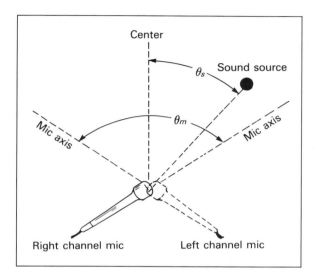

Figure 3-16
Microphone angle θ_m, source angle θ_s.

The time difference between channels is given by

$$\Delta T = \frac{\sqrt{D^2 + [(S/2) + D\tan\theta_s]^2} - \sqrt{D^2 + [(S/2) - D\tan\theta_s]^2}}{C} \qquad (2)$$

where ΔT = time difference between channels, in seconds
 D = distance from the source to the line connecting the microphones, in feet
 S = spacing between microphones, in feet
 θ_s = source angle (how far off-center the sound source is, in degrees)
 C = speed of sound (1130 ft per second)

These variables are shown in Figure 3-17.

For near-coincident microphone spacing of a few inches, the equation can be simplified to this:

$$\Delta T = \frac{S \sin \theta_S}{C} \qquad (3)$$

where ΔT = time difference between channels, in seconds
 S = microphone spacing, in inches
 θ_S = source angle, in degrees
 C = speed of sound (13,560 inches per second)

These variables are shown in Figure 3-18.

46 □ STEREO MICROPHONE TECHNIQUES

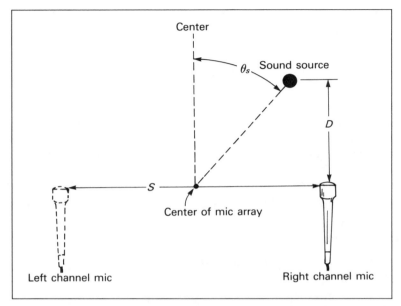

Figure 3-17 Source angle θ_s, mic-to-source distance D, and mic spacing S.

Let's consider an example. If you angle two cardioid microphones 135° apart, and the source angle is 60° (as in Figure 3-19), the dB difference produced between channels for that source is calculated as follows:

For a cardioid, $a = 0.5$ and $b = 0.5$ (from the chart following equation 1). θ_m, the angle between microphone axes, is 135°. θ_S, the source angle, is 60°. That is, the sound source is 60° off-center. So the amplitude difference between channels, using equation (1), is

Figure 3-18 Mic spacing S, and source angle θ_s.

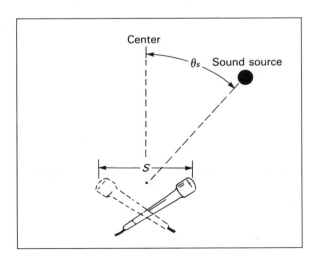

Stereo Imaging Theory □ **47**

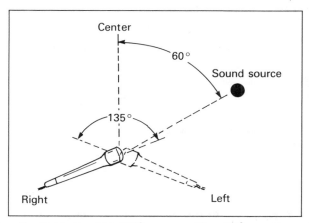

Figure 3-19 Cardioids angled 135° apart, with a 60° source angle.

$$\Delta dB = 20 \log \left[\frac{0.5 + 0.5 \cos((135°/2) - 60°)}{0.5 + 0.5 \cos((135°/2) + 60°)} \right]$$

= 14 dB amplitude difference between channels

So, according to Figure 3-10, that sound source will be reproduced nearly all the way at one speaker.

Here's another example. If you place two omnidirectional microphones 10 inches apart, and the sound source is 45° off-center, what is the time difference between channels? (Refer to Figure 3-20.)

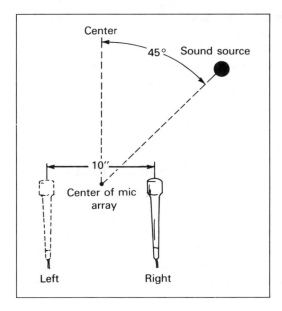

Figure 3-20 Omnis spaced 10 inches, with a 45° source angle.

Microphone spacing S is 10 inches; source angle θ_s is 45°. By equation (3),

$$\Delta T = \frac{10 \sin 45°}{13,560}$$

$$= 0.52 \text{ msec time difference between channels}$$

So, according to Figure 3-10, that sound source will be reproduced about half-way off-center.

CHOOSING ANGLING AND SPACING

There are many combinations of microphone angling and spacing that place the images of the ends of the orchestra at the right and left speaker. In other words, there are many ways to achieve a full stereo spread. You can use a narrow spacing and a wide angle, or a wide spacing and a narrow angle—whatever works. The particular angle and spacing you use is not sacred. Many do not realize this, and rely on a fixed angle and/or spacing, such as the ORTF system (110°, 17 cm). That's a good place to start, but if the reproduced stage width is too narrow, there's no harm in increasing the angle or spacing slightly.

If the center instruments are too loud, you can angle the mics farther apart while decreasing the spacing so that the reproduced stage width is unchanged. In this way, you can control the loudness of the center image to improve the balance.

To reduce pickup of early reflections from the stage floor and walls, (1) increase angling, (2) decrease spacing, and (3) place the mics closer to the ensemble. This works as follows:

1. Angling the mics farther apart softens the center instruments.
2. Decreasing the spacing between mics maintains the original reproduced stage width.
3. Since center instruments are quieter, you can place the mics closer to the ensemble and still achieve a good balance.
4. Since the mics are closer, the ratio of reflected sound to direct sound is decreased. You can add distant mics or artificial reverberation for the desired amount of hall ambience.

In general, a combination of angling and spacing (intensity and time differences) gives more accurate localization and sharper imaging than intensity or time differences alone [47].

Angling the mics farther apart increases the ratio of reverberation in the recording, which makes the orchestra sound farther away. Spacing the mics farther apart does not change the sense of distance, but it degrades the sharpness of the images.

SPACIOUSNESS AND SPATIAL EQUALIZATION

The information in this section is from Griesinger's paper, "New Perspectives on Coincident and Semi-Coincident Microphone Arrays" [48].

The *spaciousness* of a microphone array is the ratio of L−R energy to L+R energy in the reflected sound. Ideally this ratio should be equal to or greater than 1. In other words, there is equal sum and difference energy. Spaciousness implies low correlation between channels of the reflected sound.

Some microphone arrays with good spaciousness (a value of 1) are

- the spaced pair
- the Blumlein pair (figure eights crossed at 90°)
- the M−S array with a cardioid mid pattern and a 1:1 M/S ratio
- coincident hypercardioids angled 109° apart.

Spatial equalization or *shuffling* is a low-frequency shelving boost of difference (L−R) signals, and a complementary low-frequency shelving cut of sum (L+R) signals. This has two benefits:

1. It increases spaciousness, so that coincident and near-coincident arrays can sound as spacious as spaced arrays.
2. It aligns the low-frequency and high-frequency components of the sound images, which results in sharper image focus.

You can build a spatial equalizer as shown in Griesinger's paper or use an MS technique and boost the low frequencies in the L−R or side signal, and cut the low frequencies in the L+R or mid signal. The required boost/cut depends on the mic array, but a typical value is 4 to 6 dB shelving below 400 Hz. Excessive boost can split off-center images, with bass and treble at different positions. The correction should be done to the array before it is mixed with other mics.

Gerzon [49] points out that the sum and difference channels should be phase compensated (with matched, nonminimum phase responses of the two filters), as suggested by Vanderlyn [50]. Gerzon notes that spatial equalization is best applied to stereo microphone techniques not having a large antiphase reverberation component at low frequencies, such as coincident or near-coincident cardioids. With the Stereosonic technique of crossed figure eights, antiphase components tend to become excessive. He suggests a 2.4 dB cut in the sum (L+R) signal and a 5.6 dB boost in the difference (L−R) signal for better bass response.

Griesinger states, "Spatial equalization can be very helpful in coincident and semi-coincident techniques [especially when listening is done in small rooms]. Since the strongest localization information comes from the high frequencies, microphone patterns and angles can be chosen which give an accurate spread to the images at high frequencies. Spatial equalization can then be used to raise the spaciousness at low frequencies."

Alan Blumlein devised the first shuffler, revealed in his 1933 patent. He used it along with two omni mic capsules spaced apart the width of a human head. The shuffler differenced the two channels (added them in opposite polarity). When two omnis are added in opposite polarity, the result is a single bidirectional pattern aiming left and right. Blumlein used this pattern as the side pattern in an MS pair [51].

The frequency response of the synthesized bidirectional pattern is weak in the bass: it falls 6 dB/octave as frequency decreases. So Blumlein's shuffler circuit also included a first-order low-frequency boost (below 700 Hz) to compensate.

The shuffler converts phase differences into intensity differences. The farther off-center the sound source is, the greater the phase difference between the spaced mics. And the greater the phase difference, the greater the intensity difference between channels created by the shuffler.

REFERENCES

1. H. Wallach, E. Newman, and M. Rozenzweig, "The Precedence Effect in Sound Localization," *Journal of the Audio Engineering Society*, Vol. 21, No. 10, 1973 (Dec.), pp. 817–826.
2. B. Bartlett, "Stereo Microphone Technique," *db*, Vol. 13, No. 12, 1979 (Dec.), pp. 34–46.
3. M. Gardner, "Some Single and Multiple Source Localization Effects," *J. Audio Eng. Soc.*, Vol. 21, No. 6, 1973 (July/Aug.), pp. 430–437.
4. E. Carterette and M. Friedman, *Handbook of Perception*, Academic Press, New York, 1978, Vol. 4 (Hearing), pp. 62, 210.
5. P. Vanderlyn, "Auditory Cues in Stereophony," *Wireless World*, 1979 (Sept.), pp. 55–60.
6. E. Shaw, "Transformation of Sound Pressure Levels From the Free Field to the Eardrum in the Horizontal Plane," *Journal of the Acoustical Society of America*, Vol. 56, No. 6, 1974 (Dec.), pp. 1848–1861.

7. S. Mehrgardt and V. Mellert, "Transformation Characteristics of the External Human Ear," *J. Acoust. Soc. Amer.*, Vol. 61, No. 6, p. 1567.
8. F. Rumsey, *Stereo Sound for Television*, Focal Press, Boston, 1989, p. 6.
9. Rumsey, p. 3.
10. F. Harvey and M. Schroeder, "Subjective Evaluation of Factors Affecting Two-Channel Stereophony," *J. Audio Eng. Soc.*, Vol. 9, No. 1, 1961 (Jan.), pp. 19–28.
11. J. Eargle, *Sound Recording*, Van Nostrand Reinhold Company, New York, 1976, Chapters 2 and 3.
12. D. Cooper and J. Bauck, "On Acoustical Specification of Natural Stereo Imaging," *Audio Engineering Society Preprint* No. 1616 (X3), presented at the 65th Convention, 1980 February 25–28, London.
13. Rumsey, p. 4.
14. M. Gardner and R. Gardner, "Problems of Localization in the Median Plane—Effect of Pinnae Cavity Occlusion," *J. Acoust. Soc. Amer.*, Vol. 53, 1973 (Feb.), pp. 400–408.
15. Rumsey, p. 8.
16. Eargle.
17. Cooper and Bauck.
18. Ibid.
19. M. Gerzon, "Pictures of 2-Channel Directional Reproduction Systems," *Audio Engineering Society Preprint* No. 1569 (A4), presented at the 65th Convention, February 25–28, 1980, London.
20. K. Nakabayashi, "A Method of Analyzing the Quadraphonic Sound Field," *J. Audio Eng. Soc.*, Vol. 23, No. 3, 1975 (Apr.), pp. 187–193.
21. S. Koshigoe and S. Takahashi, "A Consideration on Sound Localization," *Audio Engineering Society Preprint* No. 1132 (L9), presented at the 54th Convention, May 4–7, 1976.
22. Cooper and Bauck.
23. D.H. Cooper, "Problems with Shadowless Stereo Theory: Asymptotic Spectral Status," *J. Audio Eng. Soc.*, Vol. 35, No. 9, 1987 (Sept.), p. 638.
24. C. Huggonet and J. Jouhaneau, "Comparative Spatial Transfer Function of Six Different Stereophonic Systems," *Audio Engineering Society Preprint* 2465 (H5), presented at the 82nd Convention, March 10–13, 1987, London, p. 16, Fig. 13.
25. D. Griesinger, "Equalization and Spatial Equalization of Dummy Head Recordings for Loudspeaker Reproduction," *J. Audio Eng. Soc.* Vol. 34, No. 1/2, 1989 (Jan./Feb.), pp. 20–29.
26. Eargle, pp. 122–123.
27. B. Bauer, "Stereophonic Earphones and Binaural Loudspeakers," *J. Audio Eng. Soc.*, Vol. 9, No. 2, 1961 (Apr.), pp. 148–151.
28. M. Schroeder and B. Atal, "Computer Simulation of Sound Transmission in Rooms," *IEEE Convention Record*, Part 7, 1963, pp. 150–155.
29. P. Damaske, "Head-Related Two-Channel Stereophony with Loudspeaker Reproduction," *J. Acoust. Soc. Amer.*, Vol. 50, No. 4, 1971, pp. 1109–1115.
30. T. Mori, G. Fujiki, N. Takahashi, and F. Maruyama, "Precision Sound-Image-Localization Technique Utilizing Multi-Track Tape Masters," *J. Audio Eng. Soc.*, Vol. 27, No. 1/2, 1979 (Jan./Feb.), pp. 32–38.
31. N. Sakamoto, T. Gotoh, T. Kogure, and M. Shimbo, "On the Advanced Stereophonic Reproducing System 'Ambience Stereo'," *Audio Engineering So-*

ciety Preprint No. 1361 (G3), presented at the 60th Convention, May 2–5, 1978, Los Angeles.

32. N. Sakamoto et. al., "Controlling Sound-Image Localization in Stereophonic Reproduction" (Parts I and II), *J. Audio Eng. Soc.*, Vol. 29, No. 11, 1981 (Nov.), pp. 794–799, and Vol. 30, No. 10, 1982 (Oct.), pp. 719–722.

34. H. Moller, "Reproduction of Artificial-Head Recordings through Loudspeakers," *J. Audio Eng. Soc.*, Vol. 37, No. 1/2, 1989 (Jan./Feb.), pp. 30–33.

35. D. Cooper and J. Bauck, "Prospects for Transaural Recording," *J. Audio Eng. Soc.* Vol. 37, No. 1/2, 1989 (Jan./Feb.), pp. 3–19.

36. G. Theile, "On the Stereophonic Imaging of Natural Spatial Perspective Via Loudspeakers: Theory," in *Perception of Reproduced Sound 1987*, ed. by Søren Bech and O. Juhl Pedersen, Institut für Rundfunktechnik, Floriansmuhlstrafe 60, D-8000 München 45, F.R. Germany.

37. Bartlett, pp. 38, 40.

38. E.R. Madsen, "The Application of Velocity Microphones to Stereophonic Recording," *J. Audio Eng. Soc.*, Vol. 5, No. 2, 1957 (Apr.), p. 80.

39. G. Dutton, "The Assessment of Two-Channel Stereophonic Reproduction Performance in Studio Monitor Rooms, Living Rooms, and Small Theatres," *J. Audio Eng. Soc.*, Vol. 10, No. 2, 1962 (Apr.), pp. 98–105.

40. R. Cabot, "Sound Localization in 2 and 4 Channel Systems: A Comparison of Phantom Image Prediction Equations and Experimental Data," *Audio Engineering Society Preprint* No. 1295 (J3), presented at the 58th Convention November 4–7, 1977, New York.

41. M. Williams, "Unified Theory of Microphone Systems for Stereophonic Sound Recording," *Audio Engineering Society Preprint* No. 2466 (H-6), presented at 82nd Convention, March 10–13, 1987, London.

42. J. Blauert, *Spatial Hearing*, MIT Press, Cambridge, Mass., 1983, pp. 206–207.

43. Rumsey, p. 10.

44. D. Griesinger, "New Perspectives on Coincident and Semi-Coincident Microphone Arrays," *Audio Engineering Society Preprint* No. 2464 (H4), presented at the 82nd Convention March 10–13, 1987, London.

45. Blauert.

46. Bartlett.

47. Griesinger, "New Perspectives..."

48. Ibid.

49. M. Gerzon, Letter to the Editor, Reply to comments on "Spaciousness and Localization in Listening Rooms and Their Effects on the Recording Technique," *J. Audio Eng. Soc.*, Vol. 35, No. 12, 1987 (Dec.), pp. 1013–1014.

50. P. Vanderlyn, British patent 781,186 (Aug. 14, 1957).

51. S. Lipshitz, letter to the editor, *Audio*, 1990 (April), p. 6.

4

Specific Free-Field Stereo Microphone Techniques

☐

Some stereo microphone techniques work better than others. Each method has different effects. A few techniques provide sharper imaging; some create a narrow stage width; some have exaggerated separation, and so on. In this chapter, I'll compare the characteristics of several specific stereo microphone techniques. All of these use free-field microphones; the next chapter covers stereo techniques using boundary microphones and dummy heads.

■
LOCALIZATION ACCURACY

One characteristic that varies among different types of arrays is *localization accuracy*. I'll start by explaining what this means. Localization is accurate if instruments at the sides of the ensemble are reproduced from the left or right speaker; instruments half-way off-center are reproduced half-way between the center and one speaker, and so on. In other words, there is little or no distortion of the geometry of the musical ensemble.

For example, suppose your stereo speakers are spaced the same distance apart as you're sitting from them, so that each speaker is 30° off-center. (This is the recommended arrangement for good stereo.) If the orchestral width "seen" by the microphone pair is 90°, we want sources that are 45° to one side of center to be reproduced out of only one speaker. Sources 22.5° off-center should be reproduced half-way between the center of the speaker pair and one speaker (15° off-center).

Figure 4-1 illustrates this. In Figure 4-1(a), the letters A through E represent live sound source positions relative to the microphone pair. In Figure 4-1(b), the corresponding images of these sources are accurately localized between the speaker pair.

Spacing or angling the microphones more than is necessary to achieve a full stereo spread produces an "exaggerated separation" effect: instru-

54 ☐ STEREO MICROPHONE TECHNIQUES

Figure 4-1 Stereo localization effects for a 90° (±45°) orchestral width. (a) Letters A through E represent live sound-source positions (top view). (b) Accurately localized images between speakers (listener's perception). (c) Exaggerated separation effect. (d) Narrow stage width effect.

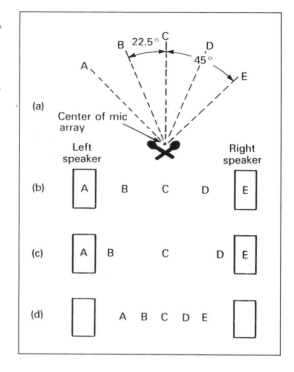

ments near the center are reproduced to the extreme left or right, rather than slightly off-center. Instruments exactly in the center are still reproduced between the speakers (see Figure 4-1(c)). Conversely, too-little angling or spacing gives a poor stereo spread or a "narrow stage width" effect (see Figure 4-1(d)).

A listening test was performed to determine the localization accuracy of various stereo microphone techniques, for a 90° orchestral width [1]. Recordings were made of a speech source at 0°, 22.5°, and 45° relative to the microphone pair (as in Figure 4-2(a)). Tests were made in an anechoic chamber and in a reverberant gymnasium. Listeners were asked to note the reproduced sound-image locations for several techniques. The image locations of the anechoic and reverberant recording rooms were averaged, with results shown in Figure 4-2(b).

Since results may vary under different listening conditions, this information is meant to be indicative, rather than definitive. Different listeners hear stereo effects differently, so your perceptions may not agree exactly with those shown. Still, Figure 4-2 lets you compare one technique to another.

The 90° orchestral width used above is arbitrary. The actual width of the orchestra varies with the size of the ensemble and the mic-to-source

Specific Free-Field Stereo Microphone Techniques □ 55

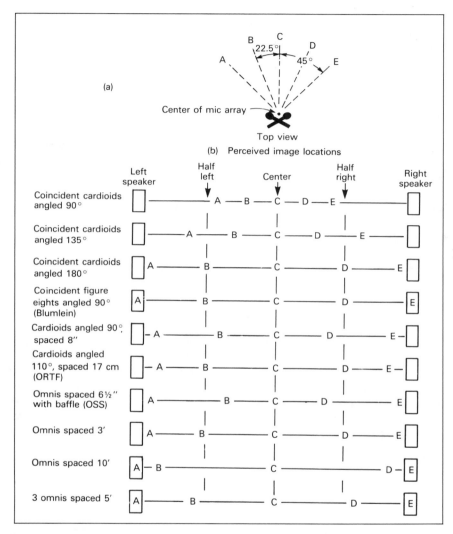

Figure 4-2 Source position versus image location of various stereo mic arrays. (a) Letters A through E are live speech-source positions relative to mic array. (b) Stereo image localization of various stereo mic arrays (listener's perception). Images A through E correspond to live speech sources A through E in (a).

distance. If the orchestral width is more than 90°, the stereo spread of all these techniques is wider than shown in Figure 4-2(b).

The closer to the ensemble a microphone array is placed, the greater is the orchestral width as seen by the microphone pair, and so, the wider is the stereo spread (up to the limit of the speaker spacing).

EXAMPLES OF COINCIDENT-PAIR TECHNIQUES

In general, coincident cardioids tend to give a narrow stereo spread and lack a sense of air or spaciousness. Imaging at high frequencies is not optimum because there is no time difference between channels, which, according to Cooper, is essential. Also, when microphones are angled apart, they receive much of the sound off axis. Many microphones have off-axis coloration (a different frequency response on and off axis).

Coincident techniques are mono-compatible: the frequency response is the same in mono and stereo. That's because there are no phase or time differences between channels to cause phase cancellations if the two channels are mixed to mono.

☐ Coincident Cardioids Angled 180° Apart

According to Figure 4-2(b), it seems reasonable to angle two coincident cardioid microphones 180° apart to achieve maximum stereo spread (as shown in Figure 4-3). However, sounds arriving from straight ahead approach each microphone 90° off axis. The 90° off-axis frequency response of some microphones is weak in high frequencies, giving a dull sound to instruments in the center of the orchestra. In addition, it has been the experience of another experimenter, Michael Gerzon, that 180° angling places the reproduced reverberation to the extreme left and right [2].

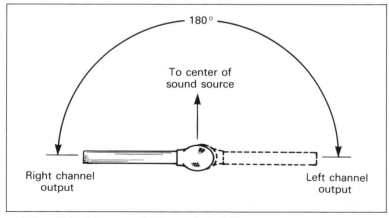

Figure 4-3   Coincident cardioids angled 180° apart.

☐ Coincident Cardioids Angled 120° to 135° Apart

A 120°-to-135° angle between microphones might be a better compromise. Gerzon has reported that the 120° angle gives a uniform spread of reverberation between speakers, while the 135° angle (Figure 4-4) provides a slightly wider stereo spread. These angles are useful where maximum stereo spread of the source is not desired. For a wider stereo spread, you can use a near-coincident or spaced pair. However, the 135° angle just described can provide a full stereo spread if the orchestral width or source angle is 150°.

☐ Coincident Cardioids Angled 90° Apart

Angling cardioids at 90° (Figure 4-5) reproduces most of the reverberation in the center. It gives a narrow stage width, unless the ensemble surrounds the microphone pair in a semicircle (180° source angle).

☐ Blumlein or Stereosonic Technique (Coincident Bidirectionals Angled 90° Apart)

This technique is diagrammed in Figure 4-6. As shown in Figure 4-2(b), it provides accurate localization. According to Gerzon [3] and the lis-

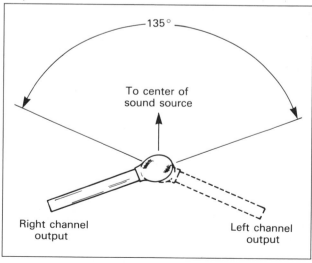

Figure 4-4   Coincident cardioids angled 135° apart.

58 □ STEREO MICROPHONE TECHNIQUES

Figure 4-5 Coincident cardioids angle 90° apart.

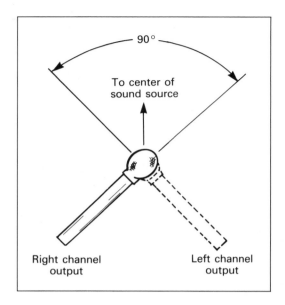

tening tests, it also provides sharp imaging, a fine sense of depth, and the most-uniform-possible spread of reverberation across the reproduced stereo stage. It has the sharpest perceived image focus of any system (other than spatially equalized systems) [4].

Note that each bidirectional pattern has a rear lobe that is in opposite polarity with the front lobe. If a sound source is more than 45° off-center (say, off to the left side), it is picked up by the front-left lobe and the back-right lobe. These are opposite polarity. This creates antiphase information between channels, which produces vague localization. For this

Figure 4-6 Blumlein or Stereosonic technique (coincident bidirectionals crossed at 90°).

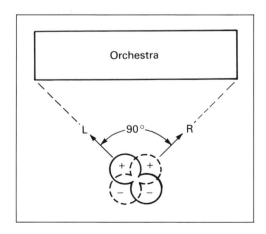

Specific Free-Field Stereo Microphone Techniques ☐ **59**

reason, the microphones should aim at the extreme left-and-right ends of the performing ensemble. This prevents sound sources from being outside the 45° limit. However, this limitation fixes the mic-to-source distance. You can't adjust this distance to vary the sense of perspective, unless you also change the angle between microphones or the size of the musical ensemble.

Another drawback is that the microphones pick up a large amount of reverberation. If you place the microphone pair closer to the ensemble to increase the direct/reverb ratio, the stereo spread becomes excessive and instruments in the center of the ensemble are emphasized. In addition, instruments at either end of the ensemble are reproduced with opposite polarity signals from both channels, so are not localized.

The Blumlein technique works best in a wide room with minimal sidewall reflections, where strong signals are not presented to the sides of the stereo pair [5].

☐ Hypercardioids Angled 110° Apart

Shown in Figure 4-7, this method give accurate localization. Listening tests also reveal sharp imaging and very good spaciousness. This array has the widest in-phase region of any array that has a spaciousness of 1 [6]. The tight pattern of the hypercardioid allows a more distant placement than with crossed cardioids. As for drawbacks, hypercardioid

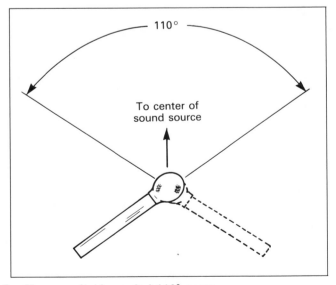

Figure 4-7 Hypercardioids angled 110° apart.

microphones tend to have a bass rolloff; but this can be corrected with equalization (bass boost).

☐ XY Shotgun Microphones

Henning Gerlach of Sennheiser Electronics suggests that two shotgun microphones can be crossed in an XY configuration with their diaphragms coincident [7]. To avoid exaggerated separation, the angle between microphones must be small. Image positioning with handheld shotguns is unstable, so this method is recommended only for stationary sound sources and stationary microphones.

Another coincident technique is mid-side (MS), which will be covered in detail later in this chapter.

EXAMPLES OF NEAR-COINCIDENT TECHNIQUES

Near-coincident techniques give wider stereo spread than coincident techniques that have the same angle between mics. In other words, if you start with a coincident technique and space the mics a few inches apart, the stereo spread will increase. Near-coincident methods also have more spaciousness and depth. This is due to the random phase relationships (low correlation) between channels at high frequencies.

These methods are not mono-compatible: if both channels are combined to mono, there are dips in the frequency response caused by phase cancellations. And since the microphones are angled apart, the sound source might be reproduced with off-axis coloration.

☐ ORTF System: Cardioids Angled 110° Apart and Spaced 17 cm (6.7") Horizontally

The listening tests summarized in Figure 4-2(b) reveal that the 110° angled, 17 cm spaced array (the ORTF, French Broadcasting Organization, system) and the 90° angled, 8-inch spaced array tend to provide accurate localization. These two methods are shown in Figures 4-8 and 4-9. According to a listening test conducted by Carl Ceoen [8], the ORTF system was preferred over several other stereo-miking techniques. It

Specific Free-Field Stereo Microphone Techniques ☐ **61**

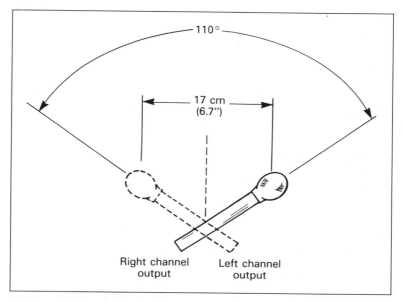

Figure 4-8 ORTF system: cardioids angled 110° and spaced 17 cm (6.7").

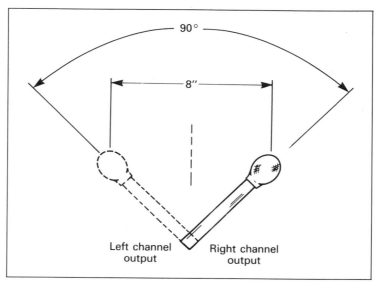

Figure 4-9 Cardioids angled 90° and spaced 8 in.

provided the best overall compromise of localization accuracy, image sharpness, an even balance across the stage, and ambient warmth.

The origin of the ORTF system was described by R. Condamines [9]. The 17 cm spacing was chosen because it provided the best image stability with head motion, assuming a speaker angle of ±30°. The 110° angle was chosen because it provided the best image precision and placement when used with a 17 cm spacing. Condamines reported that if the mic angle is less than 110°, the sound stage usually does not spread all the way between speakers; if the angle is greater than 110°, the center image becomes weak (a hole-in-the-middle effect).

The ORTF image position varies with frequency, according to calculation [10] and perception [11].

☐ NOS System: Cardioids Angled 90° Apart and Spaced 30 cm (12") Horizontally

Shown in Figure 4-10, this system was proposed by the Dutch Broadcasting Foundation. Since the spacing of the NOS system exceeds the 90° angled, 8-inch spaced array in the listening test, we could expect it to have a slightly wider stereo spread for half-left and half-right instruments.

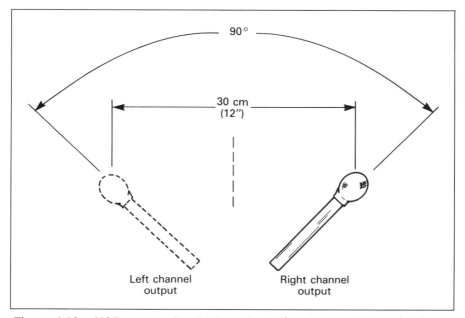

Figure 4-10 NOS system. Cardioids angled 90° and spaced 30 cm (12").

OSS (Optimal Stereo Signal or Jecklin disk)

Manufactured by Josephson Electronics, the Jecklin disk uses two omnidirectional microphones spaced 16.5 cm (6.5") and separated by a disk with a diameter of 28 cm ($11\frac{7}{8}$") [12]. The disk is hard and is covered with absorbent material to reduce reflections (Figure 4-11). The OSS system could be called quasi-binaural, in that the human binaural hearing system also uses two omni "microphones" separated by a baffle (the head).

Below 200 Hz, both microphones receive the same amplitude, and the array acts like closely spaced omnis. As frequency increases, the disk becomes more of a sound barrier, which makes the array increasingly directional. At high frequencies, the array acts like near-coincident subcardioids angled 180° apart.

Since both channels receive the same signal level at low frequencies, stereo localization at low frequencies can be due only to the capsule spacing, which causes direction-dependent delays. But, according to Griesinger [13], delay panning does not create localizable images below 500 Hz. If that's true, the OSS system localizes only above 200 Hz.

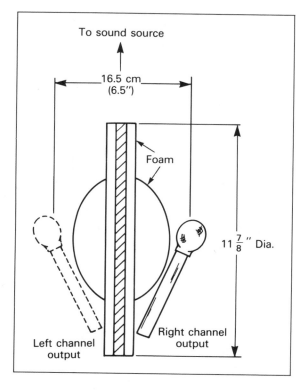

Figure 4-11 OSS system or Jecklin disk. Omnis spaced 16.5 cm (6.5") and separated by a foam-covered disk of 28 cm ($11\frac{7}{8}$") diameter.

According to the inventor, "the stereo image is nearly spectacular, and the sound is rich, full, and clear." It "seems to be superior to all other recording methods." The full sound is probably due to the use of omnidirectional condenser microphones, which have an extended low-frequency response.

Listening tests (Figure 4-2(b)) show that the OSS stereo spread for a 90° orchestral width is somewhat narrow. But, since the system uses omni microphones, it is usually placed close to the ensemble, where the angular width of the ensemble is wide. This results in a wider stereo spread. The unit is not mono-compatible.

EXAMPLES OF SPACED-PAIR TECHNIQUES

In general, listeners commented that the spaced-pair methods give relatively vague, hard-to-localize images for off-center sources. These methods are useful when you want diffuse images for special effect. Spaced arrays have a pleasing sense of spaciousness. This is produced artificially by the random phase relationships between channels, and opposite-polarity signals at various frequencies [14].

Spaced-pair techniques are not mono-compatible: peaks and dips in the frequency response of the direct sound occur when both channels are combined to mono. This effect may or may not be audible, because reverberation approaches the microphones from all angles, and each angle of sound incidence relates to a different pattern of phase cancellations. The reverberation randomizes the frequencies of these cancellations, so that the effect is less audible.

The spaced array has extreme phase differences between channels, which can make record-cutting difficult due to excess vertical modulation of the record groove by out-of-phase components. Also, the array is relatively big and unwieldy.

An advantage of the spaced-pair technique is that it allows the use of omnidirectional condenser microphones, which have a more extended low-frequency response than directional microphones. That is, the tone quality is warmer and fuller in the bass. Of course, you can equalize directional microphones to have flat bass response at a distance.

Another advantage is that the listening area for good stereo is wider than with coincident-pair techniques. The spaced-pair delay cues counteract the amplitude imbalance that occurs when the listener sits off-center.

Many instruments, such as the flute, have nulls in their radiation pattern that vary with the note played. Thus, one mic of a spaced pair might pick up a note at a low level, while the other mic would pick it up at a high level, so the image would wander with the note played. However, one mic will pick up notes that the other mic misses. Our ears have the same ability due to their spacing. Thus, the spaced pair offers the potential for better fidelity (no missed notes) at the expense of wandering images [15].

You can use cardioids or other unidirectional patterns in a spaced array to reduce pickup of hall reverberation. These patterns, however, tend to have less bass than omnis. Spaced bidirectionals have very little off-axis coloration.

☐ Omnis Spaced 3 Feet Apart

Shown in Figure 4-12, this method gives fairly accurate localization (Figure 4-2(b)), but with poorly focused imaging of off-center sources. A two-foot spacing would give more accurate localization. Since omnis must be placed relatively close to a performing ensemble for an acceptable direct/reverb ratio, this array is likely to overemphasize the center instruments. That is, the microphone pair is most sensitive to instruments in the center of the orchestra, with reduced pickup of the sides.

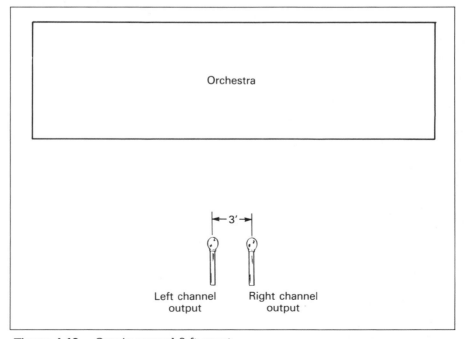

Figure 4-12 Omnis spaced 3 ft apart.

☐ Omnis Spaced 10 Feet Apart

Shown in Figure 4-13, this spacing provides a more even coverage of the orchestra (a better balance). However, spacings greater than three feet give an exaggerated separation effect, in which instruments slightly off-center are reproduced full-left or full-right (Figure 4-2(b)). This dispels the myth that spaced microphones should be as far apart as the playback loudspeakers. Instruments directly in the center of the ensemble are still reproduced exactly between the speakers.

☐ Three Omnis Spaced 5 Feet Apart (10 Feet End-to-End)

With this method (Figure 4-14), a third microphone is placed between the other two, mixed in at an approximate equal level, and split to both channels. This reduces stereo separation while maintaining full coverage of the orchestra (see Figure 4-2). The three-spaced-omnis technique is often used by Telarc Records. Image focus and mono-compatibility are fair-to-good.

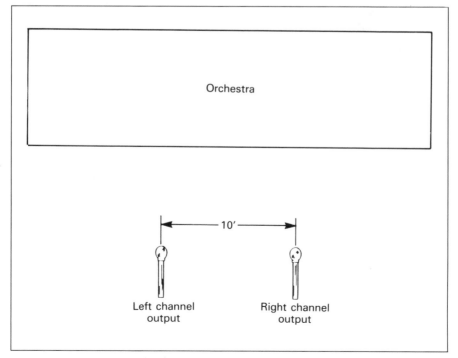

Figure 4-13   Omnis spaced 10 ft apart.

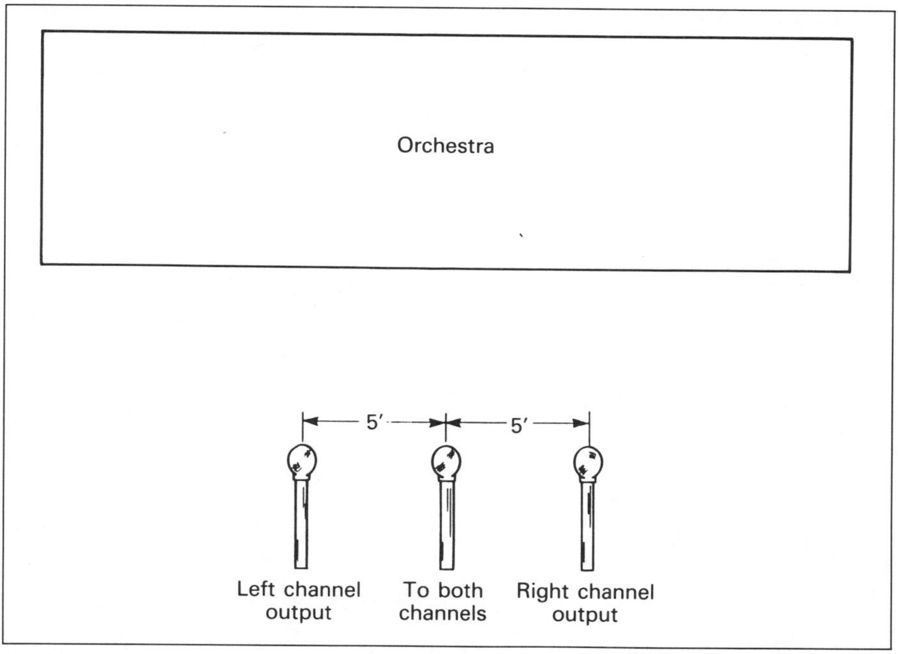

Figure 4-14 Three omnis spaced 5 ft apart.

OTHER COINCIDENT-PAIR TECHNIQUES

☐ MS (Mid-Side)

This method uses a *mid* microphone capsule aiming straight ahead toward the center of the performing ensemble, plus a side-aiming *side* bidirectional microphone capsule. These capsules are coincident and at right angles to each other (as shown in Figure 4-15.). The mid capsule is most commonly cardioid, but it can be any pattern.

The outputs of both capsules are summed to produce the left-channel signal and are differenced to produce the right-channel signal. In effect, this creates two virtual polar patterns angled apart:

$$M + S = L \quad M - S = R$$

For example, suppose that the mid capsule is omnidirectional and the side capsule is bidirectional. Also suppose that the sensitivity of both

Figure 4-15 Mid-side (MS) stereo microphone technique.

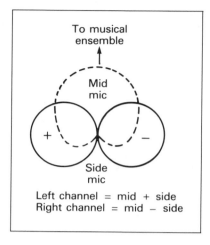

capsules is set equal. When you add these two patterns together, you get a cardioid aiming 90° to the left. When you subtract these patterns (add them in opposite polarity) you get a cardioid aiming 90° to the right. Thus, a mid-side microphone with an omni mid capsule is equivalent to two coincident cardioids angled 180° apart. A mid-side microphone with a bidirectional mid capsule is equivalent to two figure eights crossed at 90° (the Blumlein technique).

Some stereo microphones have switchable polar patterns. Changing the mid-capsule pattern changes the pattern and angling of the virtual polar patterns. The more directional the mid mic is, the more directional are the sum-and-difference (virtual) polar patterns (Figure 4-16). Consequently, you can change the apparent distance from the sound source by changing the mid pattern.

MS MATRIX BOX

The M and S outputs of the microphone are connected to an MS matrix box or decoder. This decoder uses either a tapped transformer or an active circuit (such as shown in Figure 4-17) to sum-and-difference the M and S signals. The output of the box is a left-channel signal and a right-channel signal. A schematic for an active matrix was published by Pizzi [16].

A rotating knob in the box controls the ratio of the mid signal to the side signal. By varying the ratio of mid to side, you change the polar pattern and angling of the left and right virtual mic capsules. In turn, this varies the stereo spread and the ratio of direct-to-reverberant sound. As you turn up the side signal, the stereo spread widens and the ambience increases, as shown in Figure 4-18. The optimum starting M/S ratio is near 1:1.

Specific Free-Field Stereo Microphone Techniques □ 69

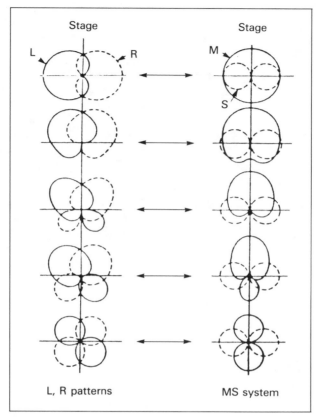

Figure 4-16 Equivalent directional patterns for MS system, with mid pattern varied. (From a letter by Les Stuck to *db* magazine, March 1981.)

By using two matrix boxes in series, you can vary the spread of any standard left/right stereo mic technique. An alternative to the matrix decoder is a mixer. As shown in Figure 4-19, you pan the M signal to the center, and split the S signal to two inputs with polarity inverted in one input. To do this, reverse the connections to pins 2 and 3 in one mic-cable connector. Pan the out-of-phase (opposite polarity) S signals hard left and right. You can vary the mid/side ratio with the microphone fader pairs.

MS ADVANTAGES

One major advantage of the MS system is that you can control the stereo spread from a remote location. This feature is especially useful for live concerts, where you can't change the microphone array during the concert. Since the stereo spread is adjustable, the MS system can be made to have accurate localization.

70 ☐ STEREO MICROPHONE TECHNIQUES

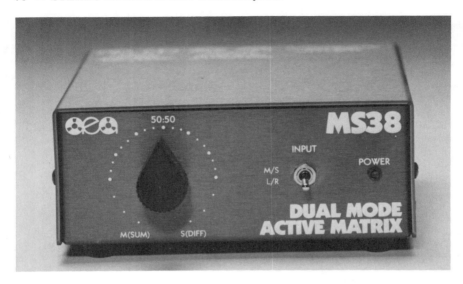

Figure 4-17 The Audio Engineering Associates MS38 active matrix decoder for MS stereo. (Courtesy of Audio Engineering Associates.)

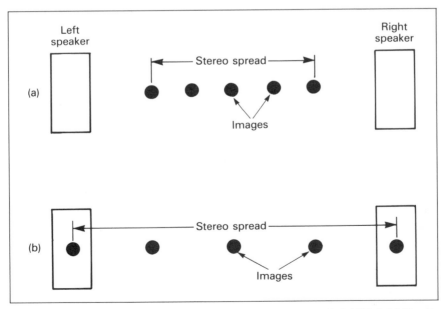

Figure 4-18 Effects of varying M/S ratio on stereo spread. (a) High M/S ratio gives a narrow spread. (b) Low M/S ratio gives a wide spread.

Specific Free-Field Stereo Microphone Techniques □ 71

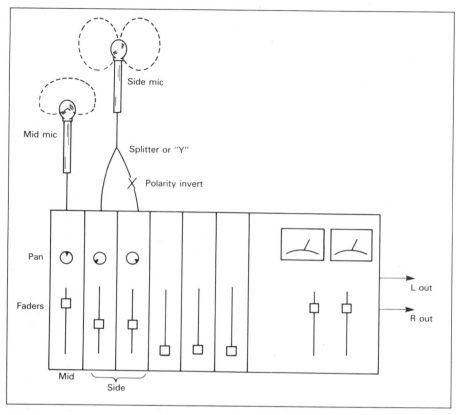

Figure 4-19 A method of using a stereo mixer as an MS matrix decoder.

If you record the M and S signals directly to a 2-track recorder during the concert, you can play them back through a matrix decoder after the concert and adjust the stereo spread then. In post-production, you can vary the spread from very narrow (mono) to very wide. While recording the concert, you monitor the outputs of the matrix decoder but do not record them.

The MS method has another advantage: it is fully mono-compatible. If you sum the left and right channels to mono, you get just the output of the forward-facing mid capsule. This is shown in the following equations:

$$\text{Left} = (M + S)$$
$$\text{Right} = (M - S)$$
$$\text{Left} + \text{Right} = (M + S) + (M - S) = 2M$$

It's as if you had a separate microphone for a mono recording. For this reason, the MS microphone sounds less reverberant in mono than in stereo, which is desirable for mono listening. And, since the mid microphone aims at the center of the orchestra, there is very little off-axis coloration compared to other coincident or near-coincident techniques.

With XY or near-coincident techniques, the center image is formed by adding the outputs of two angled directional capsules. If they are not perfectly matched in frequency response and phase response, the fusion of the center image can be degraded. But the MS system has very sharp center imaging because the center image is the output of the single mid capsule.

MS DISADVANTAGES

The MS system has been criticized for a lack of warmth, intimacy, and spaciousness [17, 18]. However, Griesinger states that MS recording can be made more spacious by giving the low frequencies a shelving boost of 4 dB (+2 dB at 600 Hz) in the L−R or side signal, with a complementary shelving cut in the L+R or mid signal.

There are other disadvantages of the MS technique. It requires the use of a matrix decoder, which is extra hardware to take on-location. If you record the MS signals on analog tape for decoding during playback, an extra tape generation is required to record the left/right decoded signals. This analog generation degrades the sound quality. A final disadvantage is that the stereo spread and direct-to-reverb ratio are interdependent: you can't change one without changing the other.

When the signals from an MS stereo microphone are mixed to mono, the resulting signal is only from the front-facing mid capsule. If this capsule's pattern is cardioid, sound sources to the far left or right will be attenuated. Thus the balance might be different in stereo and mono. If this is a problem, use an XY coincident pair rather than MS.

DOUBLE MS TECHNIQUE

Skip Pizzi recommends a *double MS* technique, which uses a close MS microphone mixed with a distant MS microphone. One MS microphone is close to the performing ensemble for clarity and sharp imaging, and the other is 50 to 75 ft out in the hall for ambience and depth. The distant mic could be replaced by an XY pair for lower cost [19].

For a comprehensive discussion of the MS system, see Streicher and Dooley [20].

MS WITH MID SHOTGUN

Henning Gerlach of Sennheiser Electronics has suggested that the MS method could be used with a shotgun microphone as the mid element [21]. He notes drawbacks of this method. The stereo spread decreases with frequency. Also, the technique is satisfactory only if the sound source is fairly close to the axis of the shotgun microphone. When you follow the main source with the shotgun, the side capsule should remain stationary or else the stereo image will shift.

Gerlach offers a way around these limitations. Use a standard MS configuration, and mix a boom-mounted, movable shotgun mic with the M signal. Use the shotgun to pick up the main source.

☐ Calrec Soundfield Microphone

This British microphone (shown in Figure 4-20) is an elaboration on the MS system. It uses four closely spaced cardioid mic capsules arranged in a tetrahedron and aiming outward. Their outputs are phase shifted to make the capsules seem perfectly coincident.

The capsule outputs are called the A-format signals. They are electronically matrixed to produce

- an omnidirectional component (the sound pressure)
- a vertical pressure-gradient component
- a left-right pressure-gradient component
- a fore-aft pressure-gradient component

These B-format signals can be further processed into stereo, quadraphonic, or ambisonic signals. Ambisonic signals include height, fore-aft, and left-right information. With a remote-control box, the user can adjust polar patterns, azimuth (horizontal rotation), elevation (vertical tilt), dominance (apparent distance), and angle (stereo spread) [22, 23].

As for drawbacks, the microphone system costs over $5000 and requires a complex matrix circuit. But it is the world's premier microphone for spatial recording.

☐ Coincident Systems with Spatial Equalization (Shuffler Circuit)

Coincident-pair systems have been criticized for a lack of spaciousness. However, as discovered by Blumlein [24], Vanderlyn [25], and Griesinger [26, 27], the focus and spaciousness can be improved by a shuffler circuit (spatial equalization). This circuit decreases stereo separation at high frequencies or increases separation at low frequencies, in order to align the image locations at low and high frequencies. To increase low-frequency separation, the circuit applies a shelving boost to low frequencies in the left-minus-right (difference) signal, and applies a complementary cut to the left-plus-right (sum) signal.

74 □ STEREO MICROPHONE TECHNIQUES

a

b

c

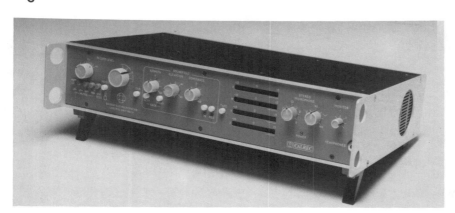

Figure 4-20 Calrec Soundfield Microphone. (a) External view. (b) Internal view, showing capsules. (c) The MK IV Control Unit. (Courtesy of AMS Industries, Inc.)

Griesinger reports that spatially equalized coincident or near-coincident arrays have very sharp imaging, and sound as spacious as a spaced array. As stated earlier, MS recordings can be made more spacious by boosting the bass 4 dB (+2 dB at 600 Hz) in the L−R or side signal, and cutting the sum signal by the same amount.

■
OTHER NEAR-COINCIDENT TECHNIQUES

☐ Stereo 180 System

Another near-coincident method is the Stereo 180 System developed by Lynn T. Olson [28], shown in Figure 4-21. It uses two hypercardioid pattern microphones, angled 135° apart, and spaced 4.6 cm (1.8") horizontally. The hypercardioid patterns have opposite-polarity rear lobes, which create the illusion that the reproduced reverberation is coming

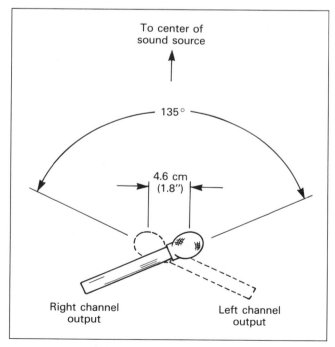

Figure 4-21 Stereo 180 system. Hypercardioids angled 135° and spaced 4.6 cm (1.8").

from the sides of the listening room as well as between the speakers. The localization accuracy and image focus of the array are reported to be very good.

☐ Faulkner Phased-Array System

Invented by Tony Faulkner, this method uses two bidirectional (figure-eight) microphones aiming straight ahead with axes parallel and spaced 20 cm (7.87") apart [29] (Figure 4-22). The plane of maximum path difference coincides with the null in the directional polar pattern of the microphones. Since the microphones are aimed forward rather than angled apart, you can place them farther from the ensemble for a better balance. This distant placement also lets you place the microphones at ear height, rather than raised. Faulkner says that the array is not mono-compatible in theory, but has presented no problems in practice.

Sometimes Faulkner adds a pair of omnidirectional microphones 2 to 3 ft apart, flanking the figure eights. These omnis add ambient spaciousness.

☐ Near-Coincident Systems with Spatial Equalization

Spatial equalization (described earlier) also improves the image focus of near-coincident methods.

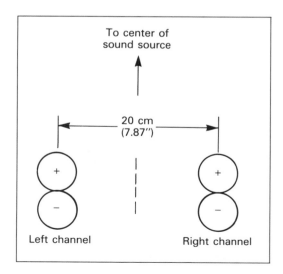

Figure 4-22 Faulkner phased-array system. Two figure eights spaced 20 cm (7.87") apart.

☐ Near-Coincident/Spaced-Pair Hybrid

John Eargle, Director of Recording at Delos International Inc., prefers to use a combination of near-coincident and spaced-pair methods [30]. A quasi-ORTF pair is placed about 4 ft behind the conductor, 9 to 11 ft high. This pair is flanked by two omnis 12 to 16 ft apart. The ORTF pair provides sharp imaging and depth, while the spaced-pair adds width to the strings and time cues from the hall. Since the spaced pair uses omnidirectional mics, low-frequency reproduction is excellent.

A second stereo pair is placed up to 30 ft behind the main pair to capture hall reverb. The woodwinds are often picked up with an overhead pair, and accent mics are added if necessary for soloists, harp, celeste, and other instruments.

COMPARISONS OF VARIOUS TECHNIQUES

Many studies have been done comparing standard stereo-miking techniques. The results of some of these are presented here. They do not all agree.

☐ Michael Williams, "Unified Theory of Microphone Systems for Stereophonic Sound Recording" [31]

Michael Williams calculated the recording angle and standard deviation of several fixed systems. *Recording angle* means the angle subtended by the sound source required for a speaker-to-speaker stereo spread. It is the angular width of the performing ensemble (as seen by the microphone array) that causes a full stereo spread.

Standard deviation means geometric distortion of the sound stage. The bigger the standard deviation in degrees, the wider is the image separation of half-left and half-right instruments. If standard deviation is zero degrees, instruments that are half-left in the orchestra are reproduced half-left between the loudspeakers (that is, at 15° off-center for speakers separated ±30°). If standard deviation is big, this is the *exaggerated-separation effect* mentioned earlier.

78 ☐ STEREO MICROPHONE TECHNIQUES

Here are his findings for various fixed mic arrays:

Coincident cardioids at 90°:
The recording angle is ±90° (180° in all). In other words, the orchestra must form a semicircle (180°) around the microphone pair to be reproduced from speaker-to-speaker.

The standard deviation is about 6°. In other words, an instrument that is half-right in the orchestra would be reproduced 6° beyond half-right.

Coincident figure eights at 90° (Blumlein):
Recording angle is ±45° (90° in all).

Standard deviation is about 5°.

Cardioids angled 110° and spaced 17 cm (ORTF):
Recording angle is ±50° (100° in all).

Standard deviation is about 5°.

Cardioids angled 90° and spaced 30 cm (NOS):
Recording angle is ±40° (80° in all).

Standard deviation is about 4°.

Omnis spaced 50 cm (20"):
Recording angle is ±50° (100° in all).

Standard deviation is about 8°.

Williams' paper has graphs showing the calculated recording angle and standard deviation for a wide range of polar patterns, anglings, and spacings, as well as other useful information.

☐ Carl Ceoen, "Comparative Stereophonic Listening Tests" [32]

Carl Ceoen used listening tests to compare several typical stereo techniques. He reported the following average resolution distortion (image focus or sharpness) for these methods:

XY (coincident cardioids angled 135°): 3°

MS (equivalent to coincident hypercardioids angled apart): 5.5°

Blumlein (coincident bidirectionals angled 90°): 4°

ORTF (coincident cardioids at 110°, 17 cm): 3°

NOS (cardioids angled 90° and spaced 30 cm): 4°

Pan-pot: 3°

According to Ceoen, the listening audience agreed that the ORTF system was the best overall compromise, and that the MS system lacked intimacy.

☐ Benjamin Bernfeld and Bennett Smith, "Computer-Aided Model of Stereophonic Systems" [33]

Bernfeld and Smith computed the image location versus frequency for various stereo-miking techniques. The better the coincidence of image locations at various frequencies, the sharper the imaging. Here are the condensed results:

Blumlein (coincident bidirectionals at 90°): Image focus is good except near the speakers; there, high frequencies are reproduced with a wider stereo spread than low frequencies.

Coincident cardioids angled 90° apart: Image focus is very good, but the stereo spread is very narrow.

Coincident cardioids angled 120° apart: Image focus is fairly good, but the stereo spread is narrow.

Coincident hypercardioids angled 120° apart: Image focus is good but not excellent because high frequencies around 3 kHz are reproduced with a wider spread than low frequencies.

Coincident hypercardioids angled 120° apart, compensated with Vanderlyn's shuffler circuit [27]: Excellent image focus and stereo spread.

Blumlein (coincident bidirectionals at 90°), compensated with shuffler circuit: Very good image focus and stereo spread.

ORTF (cardioids angled 110° and spaced 17 cm): Good image focus; low frequencies have narrow spread and high frequencies have wide spread.

ORTF with hypercardioids: Similar to the above, with wider stereo separation.

Two omnis spaced 9.5 ft: Poor image focus; high frequencies have much wider spread than low frequencies; exaggerated separation effect.

Three cardioids spaced 5 ft: Poor image focus as above, with exaggerated separation at high frequencies.

☐ C. Huggonet and J. Jouhaneau, "Comparative Spatial Transfer Function of Six Different Stereophonic Systems" [34]

Huggonet and Jouhaneau used a modulated tone burst at various frequencies, plus a violin, with listening tests to compare the spatial transfer function of six different stereophonic systems. Each system has an angular dispersion (image spread) that depends on frequency. In general, the angular dispersion of coincident systems was least. The Blumlein array gave the sharpest imaging, the dummy head and the NOS system the worst. The dummy head gave the best depth perception, followed by ORTF. MS gave the worst depth perception.

☐ M. Hibbing, "XY and MS Microphone Techniques in Comparison" [35]

In comparing XY and MS coincident methods, Hibbing concluded that MS has several advantages over XY:

1. The MS system can use an omnidirectional mid element, but the XY system cannot use omnidirectional capsules. Since an omni capsule generally has better low-frequency response than a uni, the MS system can have better low-frequency response than the XY system.
2. With MS, any stereo spread can be had with any polar pattern. XY is more limited.
3. With MS, a wider source angle is usable than with XY if polar patterns with a low bidirectional component are used.
4. With MS, the mid element aims at the center of the sound source, so most of the sound arrives close to on axis. With XY, most of the sound arrives off axis and is subject to off-axis coloration.
5. With MS, both the mid and side polar patterns are more uniform with frequency than the patterns in the XY configuration. Consequently, the left/right polar patterns generated by MS are more uniform with frequency than those of XY.
6. With MS, the stereo spread is easy to control by a fader. With XY, the stereo spread must be adjusted mechanically. MS allows stereo-spread adjustment after the session; XY does not.
7. With MS, the mid (mono sum) signal is independent of the stereo spread, so it stays consistent and predictable. With XY, the mono sum varies with the angle between the microphones.

☐ Summary

Although these experimenters disagree in certain areas, they all agree that widely spaced microphones give poorly focused imaging and that the Blumlein technique gives sharp imaging. Blumlein and Bernfeld say that the imaging of the Blumlein array can be further sharpened with a shuffler or spatial equalizer. Ceoen's results indicate that ORTF is best, but others report less-than-optimal image focus with ORTF.

The most accurate systems for frontal stereo appear to be coincident or near-coincident arrays with spatial equalization, or dual MS arrays. The near-coincident/spaced-pair hybrid method used by Delos also works quite well.

It helps to know about all the stereo techniques to conquer the acoustic problems of various halls or to create specific effects. No particular technique is magic; you can often improve the results by changing the microphone angling and/or spacing.

REFERENCES

1. B. Bartlett, "Stereo Microphone Technique," *db*, Vol. 13, No. 12, 1979 (Dec.), pp. 34–46.
2. M. Gerzon, "Blumlein Stereo Microphone Technique," *Journal of the Audio Engineering Society*, Vol. 24, No. 11, 1976 (January/February), p. 36.
3. Ibid.
4. C. Huggonet and J. Jouhaneau, "Comparative Spatial Transfer Function of Six Different Stereophonic Systems," *Audio Engineering Society Preprint* 2465 (H5), presented at the 82nd Convention, March 10–13, 1987, London, p. 11, Fig. 8.
5. R. Streicher and W. Dooley, "Basic Stereo Microphone Perspectives—A Review," *J. Audio Eng. Soc.*, Vol. 33, No. 7/8, 1985 (July/August), pp. 548–556.
6. D. Griesinger, "New Perspectives on Coincident and Semi-coincident Microphone Arrays," *Audio Engineering Society Preprint* No. 2464 (H4), presented at the 82nd Convention, March 10–13, 1987, London.
7. H. Gerlach, "Stereo Sound Recording With Shotgun Microphones," *J. Audio Eng. Soc.*, Vol. 37, No. 10, 1989 (Oct.), pp. 832–838.
8. C. Ceoen, "Comparative Stereophonic Listening Tests," *J. Audio Eng. Soc.*, Vol. 20, No. 1, 1972 (January/February), pp. 19–27. Also in the *Stereophonic Techniques Anthology*, Audio Engineering Society, 1986.
9. R. Condamines, "La Prise De Son," in *Stereophonie*, Masson Publishers, Paris and New York, 1978.
10. B. Bernfeld and B. Smith, "Computer-Aided Model of Stereophonic Systems," *Audio Engineering Society Preprint*, p. 14.
11. Huggonet, p. 14, Fig. 11.

12. J. Jecklin, "A Different Way to Record Classical Music," *J. Audio Eng. Soc.*, Vol. 29, No. 5, 1981 (May), pp. 329–332. Also in the *Stereophonic Techniques Anthology*, Audio Engineering Society, 1986.
13. Griesinger, "New Perspectives..."
14. S. Lipshitz, "Stereo Microphone Techniques: Are the Purists Wrong?", *J. Audio Eng. Soc.* Vol. 34, No. 9, 1986 (Sept.), pp. 716–744.
15. J. Lemon, "Spacing for Fidelity," letter to *Recording Engineer/Producer*, 1989 (Sept.), p. 76.
16. S. Pizzi, "Stereo Microphone Techniques for Broadcast," *Audio Engineering Society Preprint* No. 2146 (D-3), presented at the 76th Convention, October 8–11, 1984, New York.
17. Ceoen.
18. Griesinger.
19. Pizzi.
20. R. Streicher and W. Dooley, "M-S Stereo: A Powerful Technique for Working in Stereo," *J. Audio Eng. Soc.*, Vol. 33, No. 7/8, 1985 (July/August), pp. 548–555. Also available in the *Stereophonic Techniques Anthology*, published by the Audio Engineering Society, 60 E. 42nd St., New York, NY 10165.
21. Gerlach.
22. Streicher and Dooley.
23. K. Farrar, "Sound Field Microphone," *Wireless World*, October 1979.
24. A. Blumlein, British patent specification, *J. Audio Eng. Soc.*, Vol. 6, No. 2, 1958 (April), p. 91. Also in the *Stereophonic Techniques Anthology*, published by the Audio Engineering Society.
25. P. Vanderlyn, British patent specification 23989 (1954).
26. D. Griesinger, "Spaciousness and Localization in Listening Rooms and Their Effects on Recording Technique," *J. Audio Eng. Soc.*, Vol. 34, No. 4, 1986 (April), pp. 255–268.
27. Griesinger, "New Perspectives."
28. L. Olson, "The Stereo-180 Microphone System," *J. Audio Eng. Soc.*, Vol. 27, No. 3, 1979 (March), pp. 158–163. Also available in the *Stereophonic Techniques Anthology*, published by the Audio Engineering Society.
29. T. Faulkner, "Phased Array Recording," *The Audio Amateur*, January 1982.
30. *The Symphonic Sound Stage, Vol. 2*, Delos compact disc D/CD 3504.
31. M. Williams, "Unified Theory of Microphone Systems or Stereophonic Sound Recording," *Audio Engineering Society Preprint* No. 2466 (H-6), presented at the 82nd Convention, March 10–13, 1987, London.
32. Ceoen.
33. Bernfeld.
34. Huggonet and Jouhaneau.
35. M. Hibbing, "XY and MS Microphone Techniques in Comparison," *J. Audio Eng. Soc.*, Vol. 37, No. 10, 1989 (Oct.), pp. 823–831.

A mid-side computer program by John Woram is available from Gotham Audio. You input the mid pattern and mid/side ratio, and the program displays the resulting left-right polar patterns, pickup angle, and other useful information. (Gotham Audio Corp., 1790 Broadway, New York, NY 10019-1412. Tel. (212) 765-3410.)

5

Stereo Boundary-Microphone Arrays

☐

Boundary microphones (discussed in Chapter 1) can make excellent stereo recordings. This chapter explains the characteristics of several boundary-mic arrays.

First, let's look at ways to create basic stereo arrays using boundary microphones.

- To make a spaced-pair boundary array, space two boundary microphones a few feet apart. Place them on the floor, on a wall, or on stand-mounted panels.
- To make a coincident array, mount two boundary mics back-to-back on a large panel, with the edge of the panel aiming at the sound source.
- To make a near-coincident array, mount each boundary microphone on a separate panel, and angle the panels apart. Or use two directional boundary mics on the floor, angled and spaced.

Boundary microphones can be placed directly on the floor or can be raised above the floor. We'll explain several stereo techniques using both methods.

FLOOR-MOUNTED TECHNIQUES

You can place two boundary microphones on the floor to record in stereo. Floor mounting provides several advantages:

- Phase cancellations due to sound reflections off the floor are eliminated.
- Floor mounting provides the best low-frequency response for boundary microphones.

84 ☐ STEREO MICROPHONE TECHNIQUES

- The mics are very easy to place.
- The mics are nearly invisible. At live concerts, hiding the microphones is often the main consideration!

When a floor-placed boundary array is used to record an orchestra, the front-row musicians are usually reproduced too loudly, due to their relative proximity to the microphones. Musical groups with little front-to-back depth—such as small chamber groups, jazz groups, or soloists—may be the best applicaton for this system.

Let's consider specific floor-mounted techniques.

☐ Floor-Mounted Boundary Microphones Spaced 4 Feet Apart

Listening tests showed that a spacing of 3 to 4 ft between microphones is sufficient for a full stereo spread, when the sides of the musical ensemble are 45° off-center, from the viewpoint of the center of the microphone array (see Figure 5-1).

With a floor-placed array, the stereo spread decreases as the sound-source height increases. For example, if you record a group of people standing, the spread will be narrower than if the people were sitting. That's because the higher the source is, the less is the time difference between microphones.

Spaced boundary mics have the same drawbacks as spaced conventional mics: poorly focused images, potential lack of mono-compatibility, and large phase differences between channels.

Two advantages, however, are a warm sense of ambience and a good stereo effect even for off-center listeners. And, with a spaced pair, you can use omnidirectional boundary microphones without plexiglass boundaries. Thus the low-frequency response is excellent and the mics are inconspicuous.

☐ Floor-Mounted Directional Boundary Microphones

Two of these can be set up as a near-coincident pair or a spaced pair. For near-coincident use, place the mics on the floor side by side and angle them apart (Figure 5-2). Adjust their angling and spacing for the desired stereo spread. This is an effective arrangement for recording stage plays or musicals. Other mics will be needed for the pit orchestra.

As shown in Figure 5-1, a floor-mounted array of supercardioid boundary mics, angled 90° and spaced 8 inches, provides a narrow stage width.

Stereo Boundary-Microphone Arrays □ 85

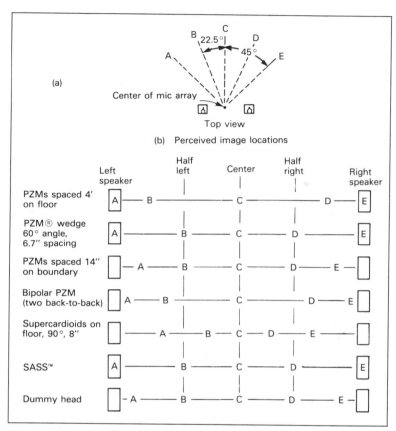

Figure 5-1 Stereo localization of various stereo boundary-microphone arrays. (a) The letters A through E are live speech-source positions relative to mic array. (b) Stereo image localization of various stereo mic arrays (listener's perception). Images A through E correspond to live speech sources A through E in (a).

More spacing or angling is needed for accurate localization. The image focus is sharper with this arrangement than with spaced boundary microphones.

□ L-Squared Floor Array

Here's a stereo PZM® array designed by Mike Lamm and John Lehmann of Dove & Note Recording, Houston, Texas [1]. It is not commercially available, but you can build one as shown in Figure 5-3. The boundaries create directional polar patterns that are angled apart and ear-spaced.

86 □ STEREO MICROPHONE TECHNIQUES

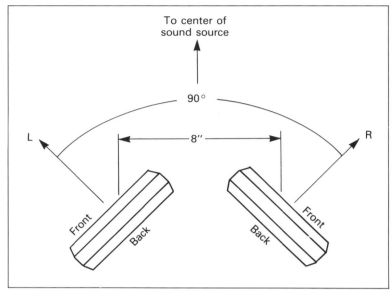

Figure 5-2 Floor-mounted directional boundary microphones set in a near-coincident array.

Figure 5-3 L^2 floor array.

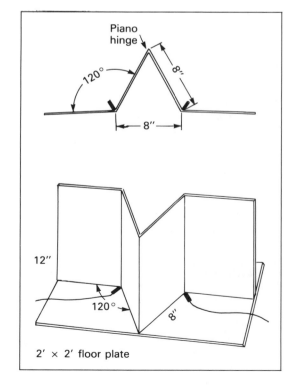

According to one user, "You can take this array, set it down, and just roll. You get a very close approximation of the real event."

Suspending the inverted array from cables results in less bass and more highs, while placing it on the floor reverses this tonal balance.

☐ OSS Boundary Microphone Floor Array

In this configuration by Josephson Engineering, two boundary microphones are on opposite sides of a hard, absorbent baffle or Jecklin disk cut in half [2]. This array has the characteristics of the OSS system described in the previous chapter, plus the advantages of boundary miking.

☐ Floor-Mounted Boundary Microphone Configured for MS

The MS technique can be applied to boundary microphones. The following method was invented by Jerry Bruck of Posthorn Recordings [3]. The mid unit is an omnidirectional boundary microphone; the side unit is a small-diameter bidirectional condenser microphone mounted a few millimeters above the omni unit.

The bidirectional microphone is close enough to the boundary to prevent phase cancellations between direct and reflected sounds over most of the audible spectrum. Bruck proposed other systems using three transducers.

Since the mid microphone is a boundary unit, it has the same high-frequency response anywhere around it (no off-axis coloration). This contributes to very sharp stereo imaging. And since the mid capsule is an omni condenser unit, it has excellent low-frequency response. The system is low-profile and unobtrusive.

Like other floor-mounted methods, this system is limited to recording small ensembles or soloists. It could also be used on a piano lid. There are no microphones made this way; you must set a bidirectional microphone over a boundary microphone to form the array.

RAISED-BOUNDARY METHODS

Some stereo microphone arrays use directional microphones. How do we make an omnidirectional boundary microphone directional? Mount it on a panel (boundary). Then you can raise the mic/panel several feet off the floor to record large ensembles. The panel makes the microphone reject sounds coming from behind the boundary. For sounds approaching the rear of the panel, low frequencies are rejected least and high frequencies are rejected most.

A small boundary makes the microphone directional only at high frequencies. Low frequencies diffract or bend around a small boundary as if it isn't there. The bigger you can make the boundary assembly, the more directional the microphone will be across the audible band.

The bigger the boundary, the lower the frequency at which the microphone becomes directional. A microphone on a square panel is omnidirectional at very low frequencies and starts to become directional above the frequency F, where

$$F = 188/D$$

D = the boundary dimension in feet

Boundaries create different polar patterns at different frequencies. For example, a 2-ft-square panel is omnidirectional at and below 94 Hz. At mid frequencies, the polar pattern becomes supercardioid. At high frequencies, the polar pattern approaches a hemisphere (as in Figure 5-4) [4].

☐ PZM® Wedge

A popular boundary configuration for stereo is the PZM® wedge [5]. You start with two clear plexiglass panels about two feet square. Join their

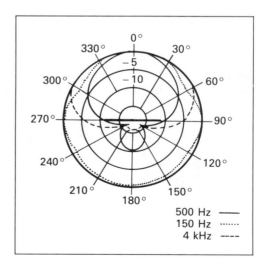

Figure 5-4 Polar patterns at various frequencies of a 2-ft square panel.

edges with a hinge or tape to form a V. Tape a PZM to each panel, as shown in Figure 5-5, and aim the point of the V at the sound source. This forms a near-coincident array which has very sharp imaging and accurate localization [6]. It also is mono-compatible to a large degree. The angle between boundaries can be varied to change the direct/reverb ratio. The wider the angle, the more forward is the directionality, and the closer the source sounds.

☐ L-Squared Array

Shown in Figure 5-6, this multipurpose array was designed by Mike Lamm and John Lehmann of Dove and Note Recording in Houston, Texas. Although it is not commercially available, you can build one as suggested in reference [7]. Mike has used this array extensively for overall stereo or quad pickup of large musical ensembles. The hinged, sliding panels can be adjusted to obtain almost any stereo pickup pattern.

☐ Pillon PZM® Stereo Shotgun

This stereo PZM array was devised by Gary Pillon, a sound mixer at General Television Network of Oak Park, Michigan [8]. Each PZM capsule is in the apex of a pyramid-shaped boundary structure; this produces a very directional polar pattern. This device is not commercially available,

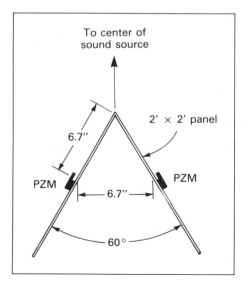

Figure 5-5 PZM® wedge.

Figure 5-6 L² array.

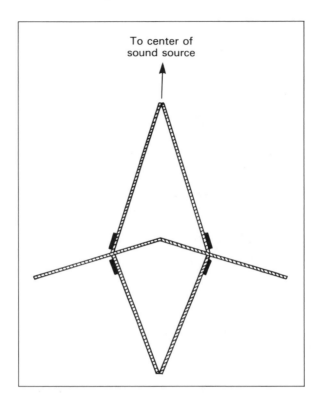

but you can build one as shown in Figure 5-7. The assembly can be stand-mounted from the backside or hand-held if necessary. The stereo imaging, which is partly a result of the 8 inch capsule spacing, is designed to be like that produced by a binaural recording, but with more realistic playback over loudspeakers. Ideally, this device would mount on a Steadicam platform and give an excellent match between audio and video perspectives.

☐ The Stereo Ambient Sampling System™

A new stereo microphone has been developed that is purported to solve many problems of other stereo-miking methods. Called the Stereo Ambient Sampling System™(SASS)™, it is a stereo condenser microphone using boundary-microphone technology [9–14]. It is designed to give highly localized stereo imaging for loudspeaker or headphone reproduction. The device is a mono-compatible, near-coincident array.

Since the microphone has an unusual design and is an example of recent technology, it requires some discussion.

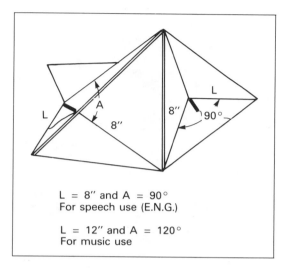

Figure 5-7 Pillon PZM® Stereo shotgun.

SASS CONSTRUCTION

One model uses two high-quality Pressure Zone Microphones mounted on boundaries to make each microphone directional (as shown in Figure 5-8). Another model is similar but uses two flush-mounted Bruel & Kjaer 4006 microphones for 10 dB lower noise.

For each channel, an omnidirectional microphone capsule is mounted very near, or flush with, a boundary approximately 12.7 cm (5") square. The two boundaries are angled left and right of center. The sound diffraction of each boundary, in conjunction with a foam barrier between the capsules, creates a directional polar pattern at high frequencies. The patterns aim left and right of center, much like a near-coincident array. The capsules are "ear-spaced" 17 cm (6.7") apart.

The polar patterns of the boundaries and the spacing between capsules have been chosen to provide natural perceived stereo imaging.

The foam barrier or baffle between the capsules limits acoustic crosstalk between the two sides at higher frequencies. Although the microphone capsules are spaced apart, there is little phase cancellation when both channels are combined to mono because of the shadowing effect of the baffle. That is, even though there are phase differences between channels, there are extreme amplitude differences (caused by the baffle) that reduce phase cancellations in mono.

SASS FREQUENCY RESPONSE

You might expect the SASS to have poor low-frequency response because it has small boundaries. However, it still has a flat response down to low frequencies; there is no 6-dB shelf. Here's why: since the capsules

92 □ STEREO MICROPHONE TECHNIQUES

Figure 5-8 Crown SASS™-P PZM® stereo microphone. (Courtesy of Crown International, Inc.)

are omnidirectional below 500 Hz, their outputs at low frequencies are equal in level. These equal-level outputs are summed in stereo listening, which causes a 3-dB rise in perceived level at low frequencies. This effectively counteracts 3 dB of the 6-dB low-frequency shelf normally experienced with small boundaries.

In addition, when the microphone is used in a reverberant sound field, the effective low-frequency level is boosted another 3 dB because the pattern is omnidirectional at low frequencies and unidirectional at high frequencies.

All of the low-frequency shelf is compensated, so the effective frequency response is uniform from 20 Hz to 20 kHz. According to the manufacturer, this can be proven in an A-B listening test by comparing the tonal balance of the SASS to that of flat-response omnidirectional microphones. They sound tonally the same at low frequencies.

SASS LOCALIZATION MECHANISMS

Like an artificial head (described in the next chapter), the SASS localizes images by time and spectral differences between channels. The localization mechanism varies with frequency:

Stereo Boundary-Microphone Arrays □ 93

- Below 500 Hz, the SASS picks up sounds equally in both channels, but with a direction-dependent delay between channels.
- Above 500 Hz, SASS localization is due to a combination of time and intensity differences. The intensity difference increases with frequency.

Although this is the opposite of the Cooper-Bauck theoretical criteria for natural imaging over loudspeakers, it seems to work well in practice. Also, it is very close to the mechanism used for binaural recording and by the human hearing system. As stated in Chapter 3, Theile suggests that head-related signals are the best for stereo reproduction.

Listening tests have shown that the SASS does not localize fundamentals with a smoothly rising and falling envelope below 261 Hz. However, the SASS uses the ears' natural ability to localize to the harmonics of a source and ignore the fundamentals. Since the SASS uses only delay panning at low frequencies, its low-frequency localization over loudspeakers might be improved by a Blumlein-type shuffler that he suggested for closely spaced omnis.

According to David Griesinger, delay panning does not work on lowpass-filtered male speech below 500 Hz [16]. However, in the book *Spatial Hearing*, Jens Blauert shows evidence by Wendt that delay panning does work at 327 Hz [17]. The effect may depend on the shape of the signal envelope, individual listening ability, and training.

Extensive application notes for the SASS are given in references [9–15].

SASS ADVANTAGES AND DISADVANTAGES

The SASS is claimed to have several characteristics that make it superior to conventional stereo microphone arrays:

- Compared to a coincident pair, the SASS has better low- frequency response and more "air" or spaciousness.
- Compared to a stereo microphone, the SASS costs much less but is relatively large.
- Compared to a near-coincident pair, the SASS has better low-frequency response. Also, it is equally mono-compatible, When you talk into the SASS from all directions in a reverberant room, it sounds tonally the same whether you listen in stereo or mono. That is, it has approximately the same frequency response in mono and stereo.
- Compared to a spaced pair, the SASS has much sharper imaging and less phase difference between channels, making it easier to cut records from SASS recordings. Also, the SASS on a single mic stand is smaller and easier to position than conventional mics on two or three stands.
- Compared to an artificial head used for binaural recording, the SASS is less conspicuous, provides a much flatter response without equalization, provides some forward directionality, and is more mono-

compatible. Its binaural localization is not quite as good as that of the artificial head.

The SASS could be called a quasi-binaural system in that it uses many of the same localization mechanisms as an artificial head. A full explanation of binaural recording is in the next chapter.

REFERENCES

1. B. Bartlett, *Boundary Microphone Application Guide*, Crown International, 1718 W. Mishawaka Rd., Elkhart, IN 46517.
2. *Josephson Engineering Catalog*, Josephson Engineering, 3729 Corkerhill, San Jose, CA 95121.
3. J. Bruck, "The Boundary Layer Mid/Side (M/S) Microphone: A New Tool," *Audio Engineering Society Preprint* No. 2313 (C-11), presented at the 79th Convention, October 12–16, 1985, New York.
4. Bartlett.
5. Ibid.
6. A. Defossez, "Stereophonic Pickup System Using Baffled Pressure Microphones," *Audio Engineering Society Preprint* No. 2352 (D4), presented at the 80th Convention, March 4–7, 1986, Montreux, Switzerland.
7. M. Lamm and J. Lehmann, "Realistic Stereo Miking for Classical Recording," *Recording Engineer/Producer*, 1983 (Aug.), pp. 107–109.
8. Bartlett.
9. B. Bartlett, "An Improved Stereo Microphone Array Using Boundary Technology: Theoretical Aspects," *Audio Engineering Society Preprint* No. 2788 (A-1), presented at the 86th Convention, March 7–10, 1989, Hamburg.
10. M. Billingsley, "Practical Field Recording Applications for An Improved Stereo Microphone Array Using Boundary Technology," *Audio Engineering Society Preprint* No. 2788 (A-1), presented at the 86th Convention, March 7–10, 1989, Hamburg.
11. M. Billingsley, "An Improved Stereo Microphone Array for Pop Music Recording," *Audio Engineering Society Preprint* No. 2791 (A-2), presented at the 86th Convention, March 7–10, 1989, Hamburg.
12. M. Billingsley, "A Stereo Microphone for Contemporary Recording," *Recording Engineer/Producer*, 1989 (November).
13. *Crown SASS Owner's Manual*, Crown International, 1718 W. Mishawaka Rd., Elkhart, IN 46517.
14. U.S. Patent 4,658,931, M. Billingsley, Apr. 21, 1987.
15. M. Billingsley, "Theory and Application of a New Near-Coincident Stereo Microphone Array for Soundtrack, Special Effects and Ambience," presented at the 89th Audio Engineering Convention, September 21–25, 1990, Los Angeles.
16. D. Griesinger, "New Perspectives on Coincident and Semi-coincident Microphone Arrays," *Audio Engineering Society Preprint* No. 2464 (H4), presented at the 82nd Convention, March 10–13, 1987, London.
17. J. Blauert, *Spatial Hearing*, MIT Press, Cambridge, MA, 1983, pp. 206–207.

I highly recommend the following recording, which demonstrates the imaging differences among various free-field stereo microphone techniques: *The Performance Recordings Demonstration of Stereo Microphone Technique* (PR-6-CD), recorded by James Boyk, Mark Fischman, Greg Jensen, and Bruce Miller; distributed by Harmonia Mundi USA, 3364 S. Robertson Blvd., Los Angeles, CA 90034, tel. (213) 559-0802. It is available in record stores nationwide.

Surprisingly, you'll hear that different transducer types have different imaging. Why? For sharpest imaging, microphone polar patterns and off-axis phase shift should be uniform with frequency. In a ribbon mic, these needs are met. But a condenser mic tends to be less uniform with frequency, and a dynamic tends to be still less uniform. These characteristics affect the imaging of a stereo pair of microphones.

6

Binaural and Transaural Techniques

☐

This chapter covers binaural recording with an artificial head (dummy head). The head contains a microphone flush mounted in each ear. You record with these microphones and play back the recording over headphones. This process can recreate the locations of the original performers and their acoustic environment with exciting realism.

Also covered in this chapter is transaural stereo, which is loudspeaker playback of binaural recordings, specially processed to provide surround-sound with only two speakers up front.

BINAURAL RECORDING AND THE ARTIFICIAL HEAD

Binaural (two-ear) recording starts with an artificial head or *dummy head*. This is a model of a human head with a flush-mounted microphone in each ear (Figure 6-1). These microphones capture the sound arriving at each ear. The microphones' signals are recorded. When this recording is played back over headphones, your ears hear the signals that originally appeared at the dummy head's ears (Figure 6-2); that is, the original sound at each ear is reproduced [1–6].

Binaural recording works on the following premise. When we listen to a natural sound source in any direction, the input to our ears is just two one-dimensional signals: the sound pressures at the ear drums. If we can recreate the same pressures at the listener's eardrums as would have occurred "live," we can reproduce the original listening experience, including directional information and reverberation [7].

Binaural recording with headphone playback is the most spatially accurate method now known. The re-creation of sound-source locations

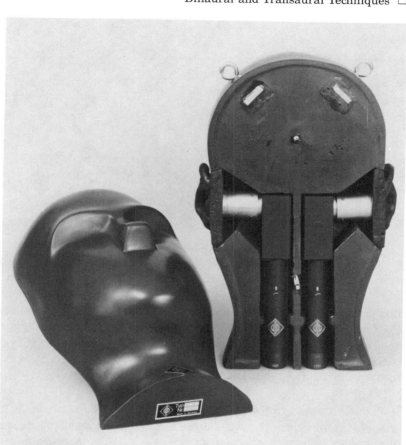

Figure 6-1 A dummy head used for binaural recording. (Courtesy of Gotham Audio Corporation.)

and room ambience is startling. Often, sounds can be reproduced all around your head—in front, behind, above, below, and so on. You may be fooled into thinking that you're hearing a real instrument playing in your listening room.

As for drawbacks: the artificial head is conspicuous—which limits its use for recording live concerts; it is not mono-compatible; and it is relatively expensive.

☐ How It Works

An artificial head picks up sound as a human head does. The head is an obstacle to sound waves at mid-to-high frequencies. On the side of the

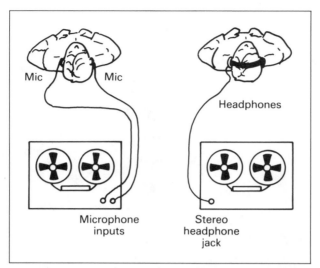

Figure 6-2 Binaural recording and headphone playback.

head away from the sound source, the ear is in a sonic shadow: the head blocks high frequencies. In contrast, on the side of the head toward the source, there is a pressure buildup (a rise in the frequency response) at mid-to-high frequencies.

The folds in the pinna (outer ear) also affect the frequency response by reflecting sounds into the ear canal. These reflections combine with the direct sound, causing phase cancellations (dips in the response) at certain frequencies.

The eardrum in a human listener is inside the ear canal, which is a resonant tube. The ear-canal's resonance does not change with sound-source direction, so the ear canal does not supply localization cues. For this reason, it is omitted in most artificial heads. Typically, the microphone diaphragm is mounted nearly flush with the head, 4 mm inside the ear canal.

To summarize: the head and outer ear cause peaks and dips in the frequency response of the sound received. These peaks and dips vary with the angle of sound incidence; they vary with the sound-source location. The frequency response of an artificial head is different in different directions. In short, the head and outer ear act as a direction-dependent equalizer.

Each ear picks up a different spectrum of amplitude and phase because one ear is shadowed by the head and the ears are spaced apart. These interaural differences vary with the source location around the head.

When the signals from the dummy-head microphones are reproduced over headphones, you hear the same interaural differences that the dummy head picked up. This creates the illusion of images located where the original sources were.

Physically, an artificial head is a near-coincident array using boundary microphones: the head is the boundary, and the microphones are flush-mounted in this boundary. The head and outer ears create directional patterns that vary with frequency. The head spaces the microphones about $6\frac{1}{2}$ inches apart. Some dummy heads include shoulders or a torso, which aids front/back localization in human listening but can degrade it in binaural recording and playback [8].

The microphones in a near-coincident array are directional at all frequencies and use no baffle between them. In contrast, the mics in an artificial head are omni at low frequencies and unidirectional at high frequencies (due to the head baffle effect).

Ideally, the artificial head is as solid as a human head, in order to attenuate sound passing through it [9]. For example, the Aachen Head is made of molded dense fiberglass [10].

You can substitute your own head for the artificial head by placing miniature condenser microphones in your ears and recording with them. The more that a dummy head and ears are shaped like your particular head and ears, the better the reproduced imaging. Thus, if you record binaurally with your own head, you might experience more precise imaging than you would if you recorded with a dummy head. This recording will have a nonflat response because of head diffraction (which I will explain later).

Another substitute for a dummy head is a head-sized sphere with flush-mounted microphones where the ears would be. This system, called the Kugelflachenmikrofon, was developed by Gunther Theile for improved imaging over loudspeakers [11].

☐ In-Head Localization

Sometimes the images are heard inside your head, rather than outside. One reason has to do with head movements. When you listen to a sound source that is outside your head and move your head slightly, you hear small changes in the arrival-time differences at your ears. This is a cue to the brain that the source is outside your head. Small movements of your head help to externalize sound sources. But the dummy head lacks this cue because it is stationary.

Here's another reason for in-head localization: the conch resonance of the pinna is disturbed by most headphones. The conch is the large cavity in the pinna just outside the ear canal. If you equalize the headphone signal to restore the conch resonance, you hear images outside the head [12].

☐ Artificial-Head Equalization

An artificial head (or a human head) has a nonflat frequency response due to the head's diffraction. This is the disturbance of a sound field by an obstacle. The diffraction of the head and pinnae create a very rough frequency response, generally with a big peak around 3 kHz for frontal sounds. Thus, binaural reproduction over headphones or loudspeakers sounds tonally colored unless custom equalization is used. Some artificial heads have built-in equalization that compensates for the effect of the head.

What is the best equalization for an artificial head to make it sound tonally like a conventional flat-response microphone? Several equalization schemes have been proposed:

- *Diffuse-field equalization.* This compensates for the head's average response to sounds arriving from all directions (such as reverberation in a concert hall).
- *Frontal free-field equalization.* This compensates for the head response to a sound source directly in front, in anechoic conditions.
- *10° averaged, free-field equalization.* This compensates for the head's response to a sound source in anechoic conditions, averaged over ±10° off center.
- *Free-field with source at ±30° equalization.* This compensates for the head's response to a sound source 30° off-center, in anechoic conditions. This is a typical stereo loudspeaker location.

The Neumann KU-81i and KEMAR artificial heads use diffuse-field equalization, which Theile also recommends. However, Griesinger found that the KU-81i needed additional equalization to sound like a Calrec Soundfield microphone: approximately +7 dB at 3 kHz and +4 dB at 15 kHz. He prefers either this equalization or a 10° averaged free-field response for artificial heads [13]. The Aachen Head, developed by Gierlich and Genuit, is equalized flat for free-field sounds in front [14], while Cooper recommends that artificial heads be equalized flat for free-field sounds at ±30° [15].

To provide a net flat response from microphone to listener, the artificial-head equalization should be the inverse of the headphone frequency response. If the head is equalized with a dip around 3 kHz to yield a net flat response, the headphones should have a mirror-image peak around 3 kHz.

☐ Artificial-Head Imaging with Loudspeakers

How does an artificial-head recording sound when reproduced over loudspeakers? According to Griesinger [16], it can sound just as good as an ordinary stereo recording, with superior reproduction of location, height, depth, and hall ambience. But it sounds even better over headphones. Images in binaural recordings are mainly up front when you listen with speakers, but are all around when you listen with headphones.

Genuit and Bray report that more reverberation is heard over speakers than over headphones, due to a phenomenon called binaural reverberance suppression [17]. For this reason, it's important to monitor artificial-head recordings with headphones and speakers.

Griesinger notes that a dummy head must be placed relatively close to the musical ensemble to yield an adequate ratio of direct-to-reverberant sound over loudspeakers. This placement yields exaggerated stereo separation with a hole in the middle. However, the center image can be made more solid by boosting in the presence range (see Griesinger's recommended equalization above) [18].

Although a dummy-head binaural recording can provide excellent imaging over headphones, it produces inadequate spaciousness at low frequencies over loudspeakers [19] unless spatial equalization is used [20]. (Spatial equalization was discussed in Chapter 3). A low-frequency boost in the L-R difference signal of about 15 dB at 40 Hz and +1 dB at 400 Hz can improve the low-frequency separation over speakers.

TRANSAURAL STEREO

It would be ideal to hear the binaural effect without having to wear headphones. That is, we'd prefer to use loudspeakers to reproduce images all around us. Our ears need only two channels to hear surround sound, so it seems as though we should be able to produce this effect with only two speakers. We can, and this process is called transaural stereo.

Transaural stereo converts binaural signals from an artificial head into surround-sound signals played over two loudspeakers. When it is done correctly, you can hear sounds in any direction around your head, with only two loudspeakers up front [21–34]. I'll explain how it works.

☐ How Transaural Stereo Works

When we listen over headphones, the right ear hears only the right signal; the left ear hears only the left signal. But when we listen over loudspeakers, there is acoustic crosstalk around the head (as shown in Figure 6-3). The right ear hears the signal not only from the right

102 □ STEREO MICROPHONE TECHNIQUES

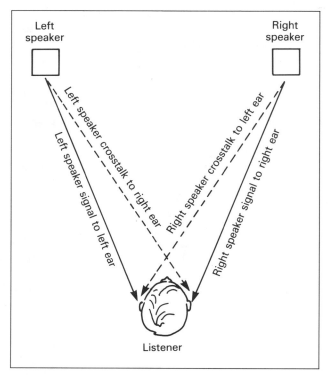

Figure 6-3 Interaural acoustic crosstalk in stereo listening.

speaker, but also the signal from the left speaker that travels around the head.

The transaural converter cancels the signal from the left speaker that reaches the right ear, and cancels the signal from the right speaker that reaches the left ear. That is, it cancels the acoustic crosstalk from each speaker to the opposite ear, so that the left ear hears only the left speaker, and the right ear hears only the right speaker. It's as if you were wearing headphones, but without the physical discomfort. The crosstalk-cancelling can be applied before or after recording.

Figure 6-4 shows a simplified block diagram of a crosstalk canceller. An equalized, delayed signal is cross-fed to the opposite channel and added in opposite polarity. This electronic crosstalk cancels out the acoustic crosstalk that occurs during loudspeaker listening.

The anti-crosstalk equalization is the difference in frequency response at the two ears due to head diffraction. The delay is the arrival-time difference between ears. The EQ and delay depend on the angle of the speakers off center, which is typically 30°.

Figure 6-5 shows how crosstalk cancellation works. Suppose we want to make a left-channel signal appear only at the left ear. That is, we want to cancel the sound from the left speaker that reaches the right ear.

Binaural and Transaural Techniques □ 103

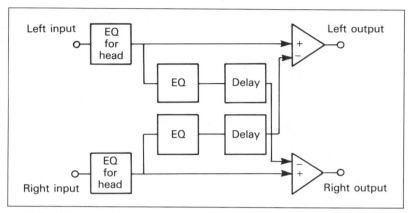

Figure 6-4 A crosstalk canceller.

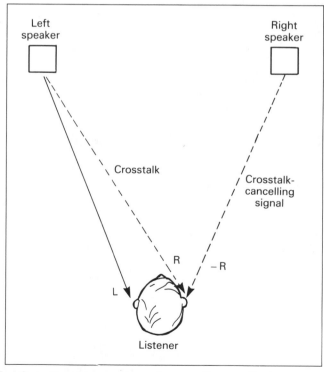

Figure 6-5 Over-the-listener view of crosstalk cancellation.

- L is the direct signal from the left speaker to the left ear.
- R is the head-diffracted (equalized, delayed) signal from the left speaker to the right ear.
- −R is the crosstalk-cancelling signal. It is an equalized, delayed, inverted version of signal L.

Signals R and −R add out-of-phase (cancel) at the right ear, so signal L is heard only at the left ear. Note that the cancellation signal itself is head-diffracted to the opposite ear, and also needs to be cancelled.

History of Transaural Stereo

Bauer was the first to suggest such a system in 1961 [35]. Atal and Schroeder first experimented with transaural stereo in 1962. They used a computer to equalize and cross-feed the two binaural channels and played the result over two speakers in an anechoic chamber. The computer program simulated complicated finite impulse response filters. If the listener moved more than a few inches from the *sweet spot* for best imaging, the effect was lost. However, Schroeder reported that the surround effect was "nothing short of amazing" [36, 37].

In the late 1970s, JVC demonstrated prototypes of a Biphonic processor that produced similar effects. JVC also researched a four-channel system called Q-Biphonic using two dummy heads, one directly in front of the other with a baffle between them [38]. Research by Matsushita led to a Ramsa sound localization control system with joysticks for surround-sound panning [39–41].

Cooper and Bauck's Crosstalk Canceller

Recently, Cooper and Bauck simplified the crosstalk-cancelling filters to minimum-phase filters in a shuffler arrangement [42]. This allowed the circuit to be realized in a few op-amps or a digital signal processing chip. They further simplified the filters at high frequencies, which enlarged the sweet spot for good imaging. In addition, they found that anechoic listening is unnecessary: all you need to do is place your speakers 1 or 2 ft from nearby walls, in order to delay early reflections at least 2 milliseconds.

Salava recommends a close-field speaker setup for transaural imaging. This monitoring arrangement provides the most accurate imaging, nearly independent of the listening room used [43].

The circuit of Cooper and Bauck generates interchannel amplitude differences and phase differences that vary with frequency. These interchannel differences are said to create interaural differences like those produced by real sound sources in various directions.

Binaural and Transaural Techniques □ 105

Basically, their circuit is made of filters and sum-and-difference sections. Filters are necessary to simulate head diffraction, because the spectrum of acoustic crosstalk around the head depends on head diffraction. Two types of filters are used independently: head-diffraction filters and anti-crosstalk filters.

The current circuit compensates for crosstalk around a sphere up to 6 kHz. Since the diffraction for a head is different than that for a sphere, you might expect different results depending on the diffraction model used. Usually the crosstalk equalization is based on the head's frequency response to a source 30° off-center, where a stereo loudspeaker is usually placed.

Their transaural circuits can be used in many ways:

- Part of a recording system to provide three-dimensional reproduction of music and concert-hall ambience. For recording, the system could use either a dummy head, a spherical stereo mic, or some other system.
- Binaural synthesis. This is an inverse shuffler that synthesizes binaural signals from a mono signal. Binaural synthesis is the inverse of crosstalk cancelling. An inverse shuffler is a crosstalk filter that mimics the crosstalk occuring when a person listens to stereo speakers.

One application of binaural synthesis is a surround-sound pan pot for recording studios. The engineer can pan a mono signal (say, from a tape track) anywhere around the listener. Another application is to process conventional microphone signals to simulate transmission around a head to the ears. This creates a binaural recording without having to use a dummy head.

A binaural synthesizer has been developed by Gierlich and Genuit [44].

- Virtual loudspeakers. These are transaural images synthesized to simulate loudspeakers placed at desired locations.

One application is a speaker spreader for stereo TVs and boom boxes: their narrow-spaced speakers can be made to sound widely spaced. The circuit could be either in each TV, in each broadcast station, or in preprocessed recordings. Boers describes a simpler speaker spreader using antiphase crosstalk [45]. Another application suggested by Cooper is a virtual center-channel speaker for theatres.

□ Lexicon's Transaural Processor

A manufactured product employing transaural principles is the Lexicon CP-1 digital audio environment processor. This consumer-type surround-sound unit can provide transaural processing of binaural recordings (among other programs). Based on work by David Griesinger, it simulates side speakers by crosstalk cancellation, and like Atal and Schroeder's method, the CP-1 cancels the signal that travels around the listener's head and also cancels the signal used for the first cancellation [46].

Griesinger reports that the sweet spot for best imaging with the CP-1 is only 2 inches wide, but that the binaural mode "provides the most realistic playback of height, depth, and surround I have yet heard through speakers."

☐ Other Surround-Sound Systems

Some other recent surround-sound methods [47–49]:

- An Ambisonics control unit, made by Audio Design. It decodes signals from the AMS Calrec Soundfield microphone (described in Chapter 4). Four speakers are needed to create the surround effect.
- Q Sound, developed by Lawrence G. Ryckman, Dan Lowe, and John Lees of Archer International Developments in Calgary, Canada, producer Jimmy Iovine and engineer Shelly Yakus. Using an A/D converter, computer, and software, the system lets the recording engineer pan mono signals up to 320° around the listener. Based on computer models of human hearing, the process is done during mixdown, and requires only two speakers without decoding during playback.
- 3-D Audio, developed by Pete Myers and Ralph Schaefer. This is produced by a large computer. The complex process, modeled on human localization mechanisms, is applied during recording.
- B.A.S.E. (Bedini Audio Spacial Environment), developed by John Bedini and marketed by Gamma Electronics. His processor separates a stereo signal into sum and difference components. The user can pan the center image forward, backward, or side to side. Although the B.A.S.E. device does not really provide surround sound, it is said to enhance stereo spaciousness and depth.

The $6000 professional unit is used only in the recording stage; the consumer unit lets the user control spatial aspects of commercial recordings.

- The Sound Retrieval System by Arnold Klayman, an engineer at Hughes Aircraft. It employs equalization to mimic the way sound enters the ear from various distances and angles. Sony recently incorporated the system into a new XBR line of stereo televisions [50].
- The Panasonic SY-DS1 Surround Sound Processor sits on a TV and produces surround sound from two amplifier/speakers built into the processor chassis.
- The Shure Surround Sound Processor, based on research by Steve Julstrom. The unit is a decoder for Dolby Stereo/Dolby Surround-encoded home videos using loudspeakers surrounding the listener [51].

- ITE™/PAR™ (In The Ear/Pinna Acoustic Response) Recording. This system was developed by Don and Carolyn Davis of Synergetic Audio Concepts. Two high-quality probe mics are inserted into the pressure zone next to the ear drum in the ear canals of a human listener. (In the future, the human listener might be replaced by a dummy head.) The microphones' signals are equalized to have a flat diffuse-field response when mounted in the ears.

During playback of the recording, you listen to four speakers: two in front (as for normal stereo) and two on either side, aimed at the ears. The speakers should be placed so as to reduce early reflections. The side speakers are said to mask opposite-ear crosstalk for a headphone-like effect. This creates a surround-sound effect from the binaural recording [52].

All these developments promise an exciting future for high-fidelity reproduction. Although stereo recording has continually been improving, it still does not put the listener in the concert hall. The reason is that all the hall reverberation is reproduced up front between the listener's pair of speakers—not all around the listener, as it is in a concert hall. While surround sound or quad can simulate reverb around the listener, it requires extra speakers and power amps. Transaural stereo does the same thing with just two speakers. It may be the biggest development in audio since digital recording.

REFERENCES

1. F. Geil, "Experiments with Binaural Recording," *db Magazine*, 1979 (June), pp. 30–35.
2. S. Peus, "Development of a New Studio Artificial Head," *db Magazine*, 1989 (June), pp. 34–36.
3. J. Sunier, "A History of Binaural Sound," *Audio*, 1989 (March), pp. 36–46.
4. J. Sunier, "Binaural Overview: Ears Where the Mics Are, Part 1," *Audio*, 1989 (Nov.), pp. 75–84.
5. J. Sunier, "Binaural Overview: Ears Where the Mics Are, Part 2," *Audio*, 1989 (Dec.), pp. 48–57.
6. K. Genuit and W. Bray, "The Aachen Head System: Binaural Recording for Headphones and Speakers," *Audio*, 1989 (Dec.), pp. 58–66.
7. H. Moller, "Reproduction of Artificial-Head Recordings through Loudspeakers," *Journal of the Audio Engineering Society*, Vol. 37, No. 1/2, 1989 (Jan./Feb.), pp. 30–33.
8. D. Griesinger, "Equalization and Spatial Equalization of Dummy Head Recordings for Loudspeaker Reproduction," *J. Audio Eng. Soc.*, Vol. 37, No. 1/2, 1989 (Jan./Feb.), pp. 20–29.
9. Sunier,... Part 2.

108 □ STEREO MICROPHONE TECHNIQUES

10. Genuit and Bray.
11. Griesinger.
12. D. Cooper and J. Bauck, "Prospects for Transaural Recording," *J. Audio Eng. Soc.*, Vol. 37, No. 1/2, 1989 (Jan./Feb.), pp. 3–19.
13. Griesinger.
14. Genuit and Bray.
15. Cooper and Bauck.
16. Griesinger.
17. Genuit and Bray.
18. Griesinger.
19. C. Huggonet and J. Jouhaneau, "Comparative Spatial Transfer Function of Six Different Stereophonic Systems." *Audio Engineering Society Preprint* No. 2465 (H5), presented at the 82nd Convention, March 10–13, 1987, London, p. 16, Fig. 13.
20. Griesinger.
21. Cooper and Bauck.
22. J. Eargle, *Sound Recording*, Van Nostrand Reinhold Company, New York, 1976, Chaps. 2 and 3.
23. B. Bauer, "Stereophonic Earphones and Binaural Loudspeakers," *J. Audio Eng. Soc.*, Vol. 9, No. 2, 1961 (Apr.), pp. 148–151.
24. M. Schroeder and B. Atal, "Computer Simulation of Sound Transmission in Rooms," *IEEE Convention Record*, Part 7, 1963, pp. 150–155.
25. T. Parsons, "Super Stereo: Wave of the Future?" *The Audio Amateur*.
26. P. Damaske, "Head-Related Two-Channel Stereophony with Loudspeaker Reproduction," *Journal of the Acoustical Society of America*, Vol. 50, No. 4, 1971, pp. 1109–1115.
27. T. Mori, G. Fujiki, N. Takahashi, and F. Maruyama, "Precision Sound-Image-Localization Technique Utilizing Multi-Track Tape Masters," *J. Audio Eng. Soc.*, Vol. 27, No. 1/2, 1979 (Jan./Feb.), pp. 32–38.
28. N. Sakamoto, T. Gotoh, T. Kogure, and M. Shimbo, "On the Advanced Stereophonic Reproducing System 'Ambience Stereo'," *Audio Engineering Society Preprint* No. 1361 (G3), presented at the 60th Convention, May 2–5, 1978, Los Angeles.
29. N. Sakamoto et. al., "Controlling Sound-Image Localization in Stereophonic Reproduction" (Parts I and II), *J. Audio Eng. Soc.*, Vol. 29, No. 11, 1981 (Nov.), pp. 794–799, and Vol. 30, No. 10, 1982 (Oct.), pp. 719–722.
30. A. Clegg, "The Shape of Things to Come: Psycho- acoustic Space Control Technology," *db*, 1979 (June), pp. 27–29.
31. K. Farrar, "Sound Field Microphone," *Wireless World*, 1979 (Oct.).
32. Moller.
33. T. Bock and D. Keele, "The Effects of Interaural Crosstalk on Stereo Reproduction and Minimizing Interaural Crosstalk in Nearfield Monitoring by the Use of a Physical Barrier, Parts 1 and 2," *Audio Engineering Society Preprint* No. 2420-A (B-10) and 2420-B (B-10), presented at the 81st Convention, November 12–16, 1986, Los Angeles.
34. Sunier,... Part 2.
35. Bauer.
36. Schroeder and Atal.
37. Parsons.
38. Mori et. al.

39. Sakomoto et. al., ref. [28].
40. Sakomoto et. al., ref. [29].
41. Clegg.
42. Cooper and Bauck.
43. T. Salava, "Transaural Stereo and Near-Field Listening," *J. Audio Eng. Soc.*, Vol. 38, No. 1/2, 1990 (Jan./Feb.), pp. 40–41.
44. Genuit and Bray.
45. P. Boers, "The Influence of Antiphase Crosstalk on the Localization Cue in Stereo Signals," *Audio Engineering Society Preprint* No. 1967 (A5), presented at the 73rd Convention, March 15–18, 1983, Eindhoven, The Netherlands.
46. Sunier,... Part 2.
47. Ibid.
48. D. Daley, "3-D Audio, Part 1," *Mix*, 1990 (Feb.), pp. 43–46.
49. B.A.S.E. literature and compact disc, Bedini Research and Development, 600 W. Broadway, Suite 100, Glendale, CA 91204.
50. *Science Digest*, 1990 (Jan.), p. 81.
51. S. Julstrom, "A High-Performance Surround Sound Process for Home Video," *J. Audio Eng. Soc.*, Vol. 35, No. 7/8, 1987 (July/Aug.), pp. 536–549.
52. *Syn-Aud-Con Newsletter*, Vol. 17, No. 1, pp. 12–13. Syn-Aud-Con, R.R. 1, Box 267, Norman, IN 47264.

At the time that this book was going into publication, several new papers on transaural stereo and surround sound were presented at the 89th convention of the Audio Engineering Society, September 21–25, 1990, Los Angeles. These papers are

"Subjective Evaluation of Spatial Image Formation Processors," Elizabeth A. Cohen, Charles M. Salter Associates, Inc., San Francisco, CA.

"An Analog LSI Dolby (TM) Pro-Logic Decoder I.C.," Peter S. Henry, Precision Monolithics, Inc., Santa Clara, CA.

"New Factors in Sound for Cinema and Television," Tomlinson Holman, University of Southern California, Los Angeles, CA, and Lucasfilm Ltd., San Rafael, CA.

"A New Method for Spatial Enhancement in Stereo and Surround Recording," Dr. Wieslaw R. Woszcyk, McGill University, Montreal, Canada.

"Multi-Channel Sound in the Home: Further Developments of Stereophony," Gunther Theile, Institut fur Rundfunktechnik, GmbH.

"Challenges to the Successful Implementation of 3-D Sound," Durand R. Begault, NASA–Ames Research Center, Moffett Field, CA.

"Directional Perception on the Cone of Confusion," William Martens, Auris Corporation, Evanston, IL.

"Digital Binaural/Stereo Conversion and Crosstalk Cancelling," Kevin Kotorynski, University of Waterloo, Ontario, Canada.

"Development and Use of Binaural Recording Technology," W. Bray, K. Genuit, and H. W. Gierlich, Jaffe Acoustics, Norwalk, CT.

"Spatial Sound Processor for Simulating Natural Acoustic Environments," Gary Kendall and Martin Wilde, Auris Corporation, Evanston, IL.

"Spaciousness Enhancement of Stereo Reproduction Using Spectral Stereo Techniques," D.J. Furlong and A. G. Garvey, Preprint 3007.

"An Intuitive View of Coincident Stereo Microphones," S. Julstrom, Preprint 2984.

A catalog of binaural recordings is available free from the *The Binaural Source*, Box 1727, Ross, CA 94957, Tel. (415) 457-9052.

7

Stereo Recording Procedures

☐

This chapter is divided into three parts. The first part explains how to do an on-location stereo recording of a classical-music ensemble [1]. The second part covers the basics of stereo miking for popular music. Finally, the last part is a special troubleshooting guide to help you pinpoint and solve problems in stereo reproduction.

Let's start by going over the equipment and procedures for an on-location recording of classical music.

EQUIPMENT

Before going on-location, you need to assemble a set of equipment, such as this:

- microphones (low-noise condenser or ribbon type, omni or directional, free field or boundary, stereo or separate)
- recorder (open-reel, DAT, Hi-Fi VCR, digital-audio adapter, etc.)
- low-noise mic preamps (unless the mic preamp in your recorder is very good)
- phantom-power supply (unless your mic preamp or mixer has phantom built-in)
- mic stands and booms or fishing line
- stereo bar
- shock mount (optional)
- microphone extension cable
- noise reduction (optional)
- mixer (optional)
- MS matrix box (optional)
- headphones and/or speakers
- power amplifier for speakers (optional)

112 □ STEREO MICROPHONE TECHNIQUES

- blank tape
- stereo phase-monitor oscilloscope (optional)
- power strip, extension cords
- notebook and pen
- tool kit

First on the list are microphones. You'll need at least two or three of the same model number or one or two stereo microphones. Good microphones are essential, for the microphones—and their placement—determine the sound of your recording. You should expect to spend $200 to $400 per microphone for professional-quality sound.

For classical-music recording, the preferred microphones are condenser or ribbon types with a wide, flat frequency response and very low self-noise (explained in Chapter 1). A self-noise spec of less than 21 dB equivalent SPL, A-weighted, is recommended.

You'll need a power supply for condenser microphones: either an external phantom-power supply, a mixer or mic preamp with phantom power, or internal batteries.

If you want to do spaced-pair recording, you can use either omnidirectional or directional microphones. Omnis are preferred because they generally have a flatter low-frequency response. If you want to do coincident or near-coincident recording for sharper imaging, use directional microphones (cardioid, supercardioid, hypercardioid, or bidirectional).

You can mount the microphones on stands or hang them from the ceiling with nylon fishing line. Stands are much easier to set up, but are more visually distracting at live concerts. Stands are more suitable for recording rehearsals or sessions with no audience present.

The mic stands should have a tripod folding base and should extend at least 14 ft high. To extend the height of regular mic stands, you can either use baby booms or use telescoping photographic stands (available from camera stores). These are lightweight and compact.

A useful accessory is a stereo bar or stereo microphone adapter. This device mounts two microphones on a single stand for stereo recording. Another needed accessory in most cases is a shock mount to prevent pickup of floor vibrations.

In difficult mounting situations, boundary microphones may come in handy. They can lie flat on the stage floor to pick up small ensembles or can be mounted on the ceiling or on the front edge of a balcony. They also can be attached to clear plexiglass panels that are hung or mounted on mic stands.

For monitoring in the same room as the musicians, you need some closed-cup, circumaural (around the ear) headphones to block out the sound of the musicians. You want to hear only what's being recorded. Of course, the headphones should be wide-range and smooth for accurate monitoring. A better monitoring arrangement might be to set up an amplifier and close-field loudspeakers in a separate room.

If you're in the same room as the musicians, you'll have to sit far from the musicians to clearly monitor what you're recording. To do that, you'll need a pair of 50-ft microphone extension cables. Longer extensions will be needed if the mics are hung from the ceiling or if you're monitoring in a separate room.

If you use noise reduction, you'll also need a small stereo microphone mixer to boost the microphones' signal level up to the line level required by the noise-reduction system. A mixer is also necessary when you want to record more than one source—for example, an orchestra and a choir, or a band and a soloist. You might put a pair of microphones on the orchestra and another pair on the choir. The mixer blends the signals of all four mics into a composite stereo signal. It also lets you control the balance (relative loudness) among microphones.

For monitoring a mid-side recording, bring an MS matrix box that converts the MS signals to L−R signals, which you monitor.

Note: be sure to test all your equipment for correct operation before going on the job.

CHOOSING THE RECORDING SITE

If possible, plan to record in a venue with good acoustics. There should be adequate reverberation time for the music being performed. This is very important, because it can make the difference between an amateur-sounding recording and a commercial-sounding one. Try to record in an auditorium, concert hall, or spacious church rather than in a band room or gymnasium.

You may be forced to record in a hall that is too dead: that is, the reverberation time is too short. In this case, you may want to add artificial reverberation from a digitial reverb unit or cover the seats with 4-mil polyethylene plastic sheeting (as Delos does).

SESSION SETUP

If the orchestral sound from the stage is bad, you might want to move the orchestra out onto the floor of the hall.

Take out your microphones and place them in the desired stereo miking arrangement. As an example, suppose you're recording an orchestra

114 □ STEREO MICROPHONE TECHNIQUES

rehearsal with two crossed cardioids on a stereo bar (the near-coincident method). Screw the stereo bar onto a mic stand and mount two cardioid microphones on the stereo bar. For starters, angle them 110° apart and space them 7 inches apart horizontally. Aim them downward so that they'll point at the orchestra when raised. You may want to mount the microphones in shock mounts or put the stands on sponges to isolate the mics from floor vibration.

Basically, you want to place two or three mics several feet in front of the group, raised up high (as in Figure 7-1). The microphone placement controls the acoustic perspective or sense of distance to the ensemble, the balance among instruments, and the stereo imaging.

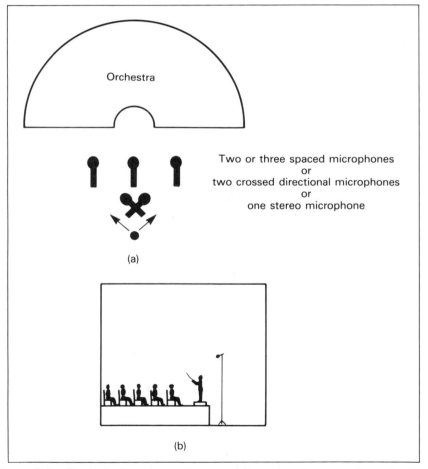

Figure 7-1 Typical microphone placement for on-location recording of a classical music ensemble. (a) Top view. (b) Side view.

As a starting position, place the mic stand behind the conductor's podium, about 12 ft in front of the front-row musicians. Connect mic cables and mic extension cords. Raise the microphones about 14 ft off the floor. This prevents overly loud pickup of the front row relative to the back row of the orchestra.

Leave some extra turns of mic cable at the base of each stand so you can reposition the stands. This slack also allows for people accidentally pulling on the cables. Try to route the mic cables where they won't be stepped on, or cover them with mats.

Live, broadcast, or filmed concerts require an inconspicuous mic placement, which may not be sonically ideal. In these cases, or for permanent installations, you'll probably want to hang the microphones from the ceiling rather than using stands. You can hang the mics by their cables or by nylon fishing line of sufficient tensile strength to support the weight of the microphones. Another inconspicuous placement is on mic-stand booms projecting forward of a balcony in front of the stage. For drama or musicals, directional boundary mics can be placed on the stage floor near the footlights.

Now you're ready to make connections. There are several different ways to do this:

- If you're using just two mics, you can plug them directly into a phantom supply (if necessary), and from there into your tape deck. You might prefer to use low-noise mic preamps, then connect cables from there into your recorder line inputs.

- If you're using two mics and a noise-reduction unit, plug the mics into a mixer to boost the mic signals up to line level, then run that line-level signal into the noise-reduction unit connected to the recorder line inputs.

- If you're using multiple mics (either spot mics or two MS mics) and a mixer, plug the mics into a snake box (described in Chapter 1). Plug the mic connectors at the other end of the snake into your mixer mic inputs. Finally, plug the mixer outputs into the recorder line inputs.

- If you're also using noise reduction, plug the mixer outputs into the inputs of the noise-reduction device and from there into the recorder.

- If you want to feed your mic signals to several mixers—for example, one for recording, one for broadcast, and one for sound reinforcement—plug your mic cables into a mic splitter or distribution amp (described in Chapter 1). Connect the splitter outputs to the snakes for each mixer. Supply phantom from one mixer only, on the microphone side of the split. Each split will have a ground-lift switch on the splitter. Set it to *ground* for only one mixer (usually the recording mixer). Set it to *lift* or *float* for the other mixers. This prevents hum caused by ground loops between the different mixers.

- If you're using directional microphones and want to make their response flat at low frequencies, you can run them through a mixer with equalization for bass boost. Boost the extreme low frequencies until the bass sounds natural or until it matches the bass response of omni condenser mics. Connect the mixer output either into an optional noise-reduction unit or directly into your recorder. This equalization will be unnecessary if the microphones have been pre-equalized by the manufacturer for flat response at a distance.

☐ Monitoring

Put on your headphones or listen over loudspeakers in a separate room. Sit equidistant from the speakers—as far from them as they are spaced apart. You'll probably need to use a close-field arrangement (speakers about 3 ft apart and 3 ft from you) to avoid coloration of the speakers' sound from the room acoustics.

Turn up the recording-level controls and monitor the signal. When the orchestra starts to play, set the recording levels to peak roughly around −10 VU so you have a clean signal to monitor. You'll set levels more carefully later on.

MICROPHONE PLACEMENT

Nothing has more effect on the production style of a classical-music recording than microphone placement. Miking distance, polar patterns, angling, spacing, and spot miking all influence the recorded sound character.

☐ Miking Distance

The microphones must be placed closer to the musicians than a good live listening position. If you place the mics out in the audience where the live sound is good, the recording will probably sound muddy and distant when played over speakers. That's because all the recorded reverberation is reproduced up-front along a line between the playback speakers, along with the direct sound of the orchestra. Close miking (5 to 20 ft from the front row) compensates for this effect by increasing the ratio of direct sound to reverberant sound.

The closer the mics are to the orchestra, the closer it sounds in the recording. If the instruments sound too close, too edgy, too detailed, or if the recording lacks hall ambience, the mics are too close to the ensemble. Move the mic stand a foot or two farther from the orchestra and listen again.

If the orchestra sounds too distant, muddy, or reverberant, the mics are too far from the ensemble. Move the mic stand a little closer to the musicians and listen again.

Eventually you'll find a *sweet spot* where the direct sound of the orchestra is in a pleasing balance with the ambience of the concert hall. Then the reproduced orchestra will sound neither too close nor too far.

Here's why miking distance affects the perceived closeness (perspective) of the musical ensemble: the level of reverberation is fairly constant throughout a room, but the level of the direct sound from the ensemble increases as you get closer to it. Close miking picks up a high ratio of direct-to-reverberant sound; distant miking picks up a low ratio. The higher the direct-to-reverb ratio, the closer the sound source is perceived to be.

An alternative to finding the sweet spot is to place a stereo pair close to the ensemble (for clarity) and another stereo pair distant from the ensemble (for ambience). According to Delos Recording Director John Eargle, the distant pair should be no more than 30 ft from the main pair. You mix the two pairs with a mixer. The advantages of this method are

- It avoids pickup of bad-sounding early reflections.
- It allows remote control (via mixer faders) of the direct/reverb ratio or the perceived distance to the ensemble.

Comb filtering due to phase cancellations between the two pairs is not severe because the delay between them is great, and their levels and spectra are different. If the distant pair is farther back than 30 feet, its signal might simulate an echo.

Skip Pizzi recommends a "double MS" technique, which uses a close MS microphone mixed with a distant MS microphone (as shown in Figure 7-2). One MS microphone is close to the performing ensemble for clarity and sharp imaging, and the other is out in the hall for ambience and depth. The distant mic could be replaced by an XY pair for lower cost [2].

If the ensemble is being amplified through a sound-reinforcement system, you might be forced to mike very close to avoid picking up amplified sound and feedback from the reinforcement speakers.

For broadcast or communications, consider miking the conductor with a wireless lavalier mic.

☐ Stereo-Spread Control

Concentrate on the stereo spread. If the monitored spread is too narrow, it means that the mics are angled or spaced too close together. Increase the angle or spacing between mics until localization is accurate.

118 ☐ STEREO MICROPHONE TECHNIQUES

Figure 7-2 Double MS technique using a close main pair and a distant pair for ambience. Spot mics are also shown.

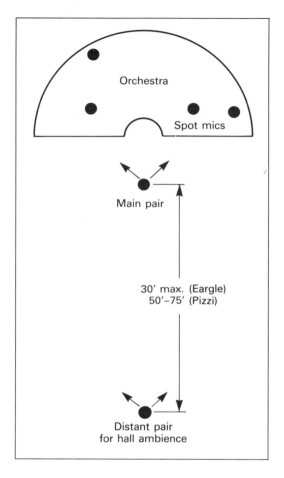

Note: increasing the angle between mics will make the instruments sound farther away; increasing the spacing will not.

If off-center instruments are heard far-left or far-right, that indicates your mics are angled or spaced too far apart. Move them closer together until localization is accurate.

If you record with a mid-side microphone, you can adjust the stereo spread by remote control at the matrix box with the stereo spread control (M/S ratio control).

You can change the monitored stereo spread either during the recording or after.

- To change the spread during the recording, connect the stereo-mic output to the matrix box and connect the matrix-box output to the recorder. Use the stereo-spread control (M/S ratio) in the matrix box to adjust the stereo spread.

- To alter the spread after the recording, record the mid signal on one track and the side signal on another track. Monitor the output of the recorder with a matrix box. After the recording, run the mid and side tracks through the matrix box, adjust the stereo spread as desired, and record the result.

If you are set up before the musicians arrive, check the localization by recording yourself speaking from various positions in front of the microphone pair while announcing your position (e.g., "left side," "mid-left," "center"). Play back the recording to judge the localization accuracy provided by your chosen stereo array. Recording this localization test at the head of a tape is an excellent practice.

MONITORING STEREO SPREAD

Full stereo spread on speakers is a spread of images all the way between speakers, from the left speaker to the right speaker. Full stereo spread on headphones can be defined as stereo spread from ear to ear. The stereo spread heard on headphones may or may not match the stereo spread heard over speakers, depending on the microphone technique used.

Due to psychoacoustic phenomena, coincident-pair recordings have less stereo spread over headphones than over loudspeakers. Take this into account when monitoring with headphones or use only loudspeakers for monitoring.

If you are monitoring your recording over headphones or anticipate headphone listening to the playback, you may want to use near-coincident miking techniques, which have similar stereo spread on headphones and loudspeakers.

Ideally, monitor speakers should be set up in a close-field arrangement (say, 3 ft from you and 3 ft apart) to reduce the influence of room acoustics and to improve stereo imaging.

If you want to use large monitor speakers placed farther away, deaden the control-room acoustics with Sonex™ or thick fiberglass insulation (covered with muslin). Place the acoustic treatment on the walls behind and to the sides of the loudspeakers. This smooths the frequency response and sharpens stereo imaging.

You'll probably want to include a stereo/mono switch in your monitoring system, as well as an oscilloscope. The 'scope is used to check for excessive phase shift between channels, which can degrade mono frequency response or cause record-cutting problems. Connect the left-channel signal to the 'scope's vertical input; connect the right-channel signal to the horizontal input, and look for the lissajous patterns shown in Figure 7-3.

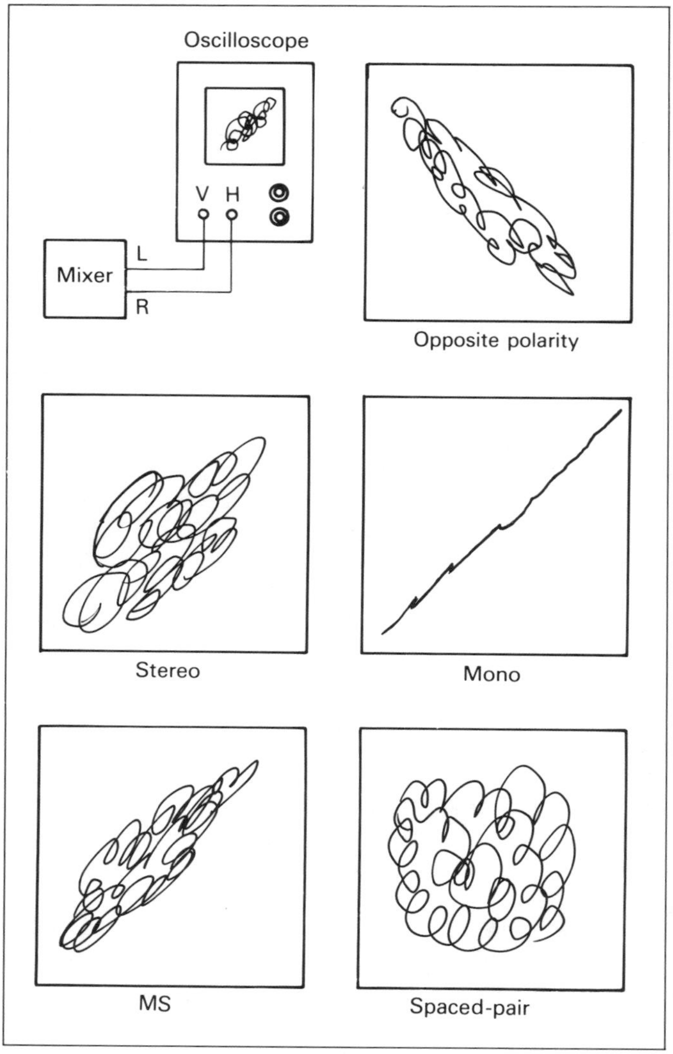

120 ☐ STEREO MICROPHONE TECHNIQUES

Figure 7-3 Oscilloscope lissajous patterns showing various phase relationships between channels of a stereo program.

☐ Soloist Pickup and Spot Microphones

Sometimes a soloist plays in front of the orchestra. You'll have to capture a tasteful balance between the soloist and the ensemble. That is, the main stereo pair should be placed so that the relative loudness of the soloist and the accompaniment is musically appropriate. If the soloist is too loud relative to the orchestra (as heard on headphones or loudspeakers), raise the mics. If the soloist is too quiet, lower the mics. You may want to add a spot mic (accent mic) about 3 ft from the soloist and mix it with the other microphones. Take care that the soloist appears at the proper depth relative to the orchestra.

Many record companies prefer to use multiple microphones and multitrack techniques when recording classical music. Such methods provide extra control of balance and definition and are necessary in difficult situations. Often you must add spot or accent mics on various instruments or instrumental sections to improve the balance or enhance clarity (as shown in Figure 7-2). In fact, John Eargle contends that a single stereo pair of mics rarely works well.

Pan each spot mic so that its image position coincides with that of the main microphone pair. Using the mute switches on your mixing console, alternately monitor the main pair and each spot to compare image positions.

You might want to use an MS microphone for each spot mic and adjust the stereo spread of each local sound source to match that reproduced by the main pair. For example, suppose that a violin section appears 20° wide as picked up by the main pair. Adjust the perceived stereo spread of the MS spot mic used on the violin section to 20°, then pan the center of the section image to the same position that it appears with the main mic pair.

When you use spot mics, mix them at a low level relative to the main pair—just loud enough to add definition, but not loud enough to destroy depth. Operate the spot-mic faders subtly or leave them untouched. Otherwise the close-miked instruments may seem to jump forward when the fader is brought up, then fall back in when the fader is brought down. If you bring up a spot-mic fader for a solo, drop it only 6 dB when the solo is over—not all the way off.

Often the timbre of the instrument(s) picked up by the spot mic is excessively bright. You can fix it with a high-frequency rolloff. Adding artificial reverb to the spot mic can help too.

To further integrate the sound of the spots with the main pair, you might want to delay each spot's signal to coincide with those of the main pair. That way, the main and spot signals are heard at the same time. For each spot mic, the formula for the required delay is

$$T = D/C$$

where T = delay time in seconds
 D = distance between each spot mic and the main pair in feet
 C = speed of sound, 1130 ft per second.

For example, if a spot mic is 20 ft in front of the main pair, the required delay is 20/1130 or 17.7 msec. Some engineers add even more delay (10–15 msec) to the spot mics to make them less noticeable [3].

☐ Setting Levels

Once the microphones are positioned properly, you're ready to set recording levels. Ask the orchestra to play the loudest part of the composition, and set the recording levels for the desired meter reading. A typical recording level is +3 VU maximum on a VU meter or −5 dB maximum on a peak-reading meter for a digital recorder.

When recording a live concert, you'll have to set the record-level knobs to a nearly correct position ahead of time. Do this during a pre-concert trial recording, or just go by experience: set the knobs where you did at previous sessions (assuming you're using the same microphones at this session).

☐ Multitrack Recording

British Decca has developed an effective recording method using an 8-track recorder [4]:

- record the main pair on two tracks
- record the distant pair on two tracks
- record panned accent mics on two tracks
- mix down the three pairs of tracks to two stereo tracks

All tracks should be recorded at full level (approximately 0 VU) with noise reduction.

STEREO MIKING FOR POP MUSIC

Most current pop music recordings are made using multiple close-up microphones (one or more on each instrument). These multiple mono sources are panned into position and balanced with faders. Such an approach is convenient but often sounds artificial. The size of each instrument is reduced to a point, and each instrument might sound isolated in its own acoustic space.

The realism can be greatly enhanced by stereo miking parts of the ensemble and overdubbing several of these stereo pickups. Such a technique can provide the feeling of a musical ensemble playing together in a common ambient space [5, 6]. Realism is improved for several reasons:

- The more-distant miking provides more-natural timbre reproduction.
- The size of each instrument is reproduced.
- Time cues for localization are included (with near-coincident and spaced techniques).
- The sound of natural room acoustics is included.

True-stereo recording works especially well for these sound sources:

- drum kit (overhead)
- piano (out front and in line with the lid, or over the strings)
- backing vocals
- horn and string sections
- vibraphone and xylophone
- other percussion instruments and ensembles

You can even stereo-mike a soloist or a singer/guitarist in an ambient space.

If you record several performers with a stereo pair, this method has some disadvantages. You must adjust their balance by moving the performers toward and away from the mics during the session. This is more time-consuming and expensive than moving faders for individual tracks after the session. In addition, the performances are not acoustically isolated. So if someone makes a mistake, you must re-record the whole ensemble rather than just the flawed performance.

The general procedures for true-stereo overdubs are

1. Adjust the acoustics around the instruments. Add padding or reflective surfaces if necessary.
2. Place the musicians around the stereo mic pair where you want them to appear in the final mix. For example, you might overdub strings spread between center and far right and horns spread between center and far left.
3. Experiment with different microphone heights (to vary the tonal balance) and miking distance (to vary the amount of ambience).
4. If some instruments or vocalists are too quiet, move them closer to the stereo mic pair, and vice versa, until the balance is satisfactory.

Some excellent application notes for pop-music stereo miking are found in references [5, 6].

TROUBLESHOOTING STEREO SOUND

Suppose that you're monitoring a recording in progress or listening to a recording you've already made. Something doesn't sound right. How can you pinpoint what's wrong and how can you fix it?

This section lists several procedures to solve audio-related problems. Read down the list of bad sound descriptions until you find one matching what you hear, then try the solutions until your problem disappears.

Before you start, check for faulty cables and connectors. Also check all control positions; rotate knobs and flip switches to clean the contacts.

☐ Distortion in the Microphone Signal

- Use pads or input attenuators in your mixer.
- Switch in the pad in the condenser microphone, if any.
- Use a microphone with a higher "Maximum SPL" specification.

☐ Too Dead (Insufficient Ambience, Hall Reverberation, or Room Acoustics)

- Place microphones farther from performers.
- Use omnidirectional microphones.
- Record in a concert hall with better acoustics (longer reverberation time).
- Add artificial reverberation.
- Add plastic sheeting over the audience seats.

☐ Too Detailed, too Close, too Edgy

- Place microphones farther from performers.
- Place microphones lower or on the floor (as with a boundary microphone).
- Using an equalizer in your mixing console, roll off the high frequencies.
- Use duller-sounding microphones.
- If using both a close-up pair and a distant ambience pair, turn up the ambience pair.

- If using spot mics, add artificial reverb or delay the signal to coincide with that of the main pair.

☐ Too Distant (too much Reverberation)

- Place microphones closer to performers.
- Use directional microphones (such as cardioids).
- Record in a concert hall that is less "live" (reverberant).
- If using both a close-up pair and a distant ambience pair, turn down the ambience pair.

☐ Narrow Stereo Spread (Figure 7-4(c))

- Angle or space the main microphone pair farther apart.
- If doing mid-side stereo recording, turn up the side output of the stereo microphone.
- Place the main microphone pair closer to the ensemble.
- If monitoring with headphones, narrow stereo spread is normal when you use coincident techniques. Try monitoring with loudspeakers, or use near-coincident or spaced techniques.

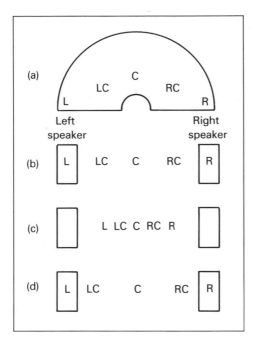

Figure 7-4 Stereo localization effects. (a) Orchestra instrument locations (top view). (b) Images accurately localized between speakers (the listener's perception). (c) Narrow stage-width effect. (d) Exaggerated separation effect.

☐ Excessive Separation or Hole-in-the-Middle (Figure 7-4(d))

- Angle or space the main microphone pair closer together.
- If doing mid-side stereo recording, turn down the side output of the stereo microphone or use a cardioid mid instead of an omni mid.
- In spaced-pair recording, add a microphone midway between the outer pair and pan its signal to the center.
- Place the microphones farther from the performers.
- Place the loudspeaker pair closer together. Ideally, they should be as far apart as you are sitting from them, to form a listening angle of 60°.

☐ Poorly Focused Images

- Avoid spaced-microphone techniques.
- Use a spatial equalizer (described in Chapter 3).
- Use a microphone pair that is better-matched in frequency response and phase response.
- If the sound source is out of the in-phase region of microphone pickup, move the source or the microphone. For example, the in-phase region of a Blumlein pair of crossed figure eights is ±45° relative to center.
- Be sure that each spot mic is panned so that its image location coincides with that of the main pair.
- Use loudspeakers designed for sharp imaging. Usually these are signal-aligned, have vertically aligned drivers, have curved edges to reduce diffraction, and are sold in carefully matched pairs.
- Place the loudspeakers several feet from the wall behind them and from side walls to delay and weaken the early reflections that can degrade stereo imaging.

☐ Images Shifted to One Side (Left-Right Balance Is Faulty)

- Adjust the right-or-left recording level so that center images are centered.
- Use a microphone pair that is better-matched in sensitivity.
- Aim the center of the mic array exactly at the center of the ensemble.
- Sit exactly between your stereo speakers, equidistant from them. Adjust the balance control or level controls on your monitor amplifier to center a mono signal.

☐ Lacks Depth (Lacks a Sense of Nearness and Farness of Various Instruments)

- Use only a single pair of microphones out front. Avoid multi-miking.
- If you must use spot mics, keep their level low in the mix, and delay their signals to coincide with those of the main pair.

☐ Lacks Spaciousness

- Use a spatial equalizer (described in Chapter 3).
- Space the microphones apart.
- Place the microphones farther from the ensemble.

☐ Early Reflections too Loud

- Place mics closer to the ensemble and add a distant microphone for reverberation (or use artificial reverberation).
- Place the musical ensemble in an area with weaker early reflections.
- If the early reflections come from the side, try aiming bidirectionals at the ensemble. Their nulls will reduce pickup of side-wall reflections.

☐ Bad Balance (Some Instruments too Loud or too Soft)

- Place the microphones higher or farther from the performers.
- Ask the conductor or performers to change the instruments' written dynamics.
- Add spot microphones close to instruments or sections needing reinforcement. Mix them in subtly with the main microphones' signals.
- Increase the angle between mics to reduce the volume of center instruments, and vice versa.
- If the center images of a mid-side recording are weak, use a cardioid mid instead of an omni mid [7].

☐ Muddy Bass

- Aim the bass-drum head at the microphones.
- Put the microphone stands and bass-drum stand on resilient isolation mounts, or place the microphones in shock-mount stand adapters.

- Roll off the low frequencies or use a highpass filter set around 40 to 80 Hz.
- Record in a concert hall with less low-frequency reverberation.

☐ Rumble from Air Conditioning, Trucks, and so on

- Temporarily shut off air conditioning.
- Record in a quieter location.
- Use a high-pass filter set around 40 to 80 Hz.
- Use microphones with limited low-frequency response.

☐ Bad Tonal Balance (too Dull, too Bright, Colored)

- Change the microphones.
- If a microphone must be placed near a hard reflective surface, use a boundary microphone to prevent phase cancellations between direct and reflected sounds.
- Adjust equalization. Compared to omni condenser mics, directional mics usually have a rolled-off low-frequency response and may need some bass boost.
- If strings sound strident, move mics farther away or lower.
- If the tone quality is colored in mono monitoring, use coincident-pair techniques.

REFERENCES

1. B. Bartlett, "Microphone-Technique Basics," and "On-Location Recording of Classical Music," in *Introduction to Professional Recording Techniques*, ed. by John Woram, Howard W. Sams & Co., Indianapolis, Chapters 7 and 17, pp. 109–117, 301–310.
2. S. Pizzi, "Stereo Microphone Techniques for Broadcast," *Audio Engineering Society Preprint* No. 2146 (D-3), presented at the 76th Convention, October 8–11, 1984, New York.
3. R. Streicher and W. Dooley, "Basic Stereo Microphone Perspectives—A Review," *J. Audio Eng. Soc.*, Vol. 33, No. 7/8, 1985 (July/Aug.), pp. 548–556.
4. J. Eargle, *The Microphone Handbook*, Elar Publishing, Plainview, New York, 1981, pp. 119–121.

5. M. Billingsley, "An Improved Stereo Microphone Array for Pop Music Recording," *Audio Engineering Society Preprint* No. 2791 (A-2), presented at the 86th Convention, March 7–10, 1989, Hamburg.
6. M. Billingsley, "A Stereo Microphone for Contemporary Recording," *Recording Engineer/ Producer*, 1989 (Nov.).

8

Broadcast, Film and Video, Sound Effects, and Sampling

☐

Stereo microphone techniques have many uses besides music recording. With the advent of stereo TV (MTS) and films using Dolby™ Stereo or THX™, the need for stereo microphone techniques has never been greater. In this chapter, we'll explore the application of stereo miking to broadcast, film and video, sound effects, and sampling. First, some tips that apply to all these areas.

For on-location work, if wind noise and mechanical vibration are problems, use a windscreen and a shock mount. An alternative is to use a stereo pair of omnidirectional microphones. These are inherently less sensitive to wind and vibration than directonal types. If you need an omni pair with sharp imaging, try one of the PZM boundary arrays shown in Chapter 5, or the OSS array shown in Chapter 4.

If you want sharp imaging and precise localization, try a coincident or near-coincident technique. For a diffuse, spacious sound (for example, background ambience), try a spaced-pair technique. A coincident pair or Crown SASS microphone is recommended when mono-compatibility is important—which is almost always!

When the signals from an MS stereo microphone are mixed to mono, the resulting signal is only from the front-facing mid capsule. If this capsule's pattern is cardioid, sound sources to the far left and right will be attenuated. Thus, the balance might be different in stereo and mono. If this is a problem, use an XY coincident pair rather than MS.

When compressing or limiting a stereo program, you run each channel's signal through a separate channel of compression. Connect or gang these two channels of compression so that they track together. Otherwise, the image positions will shift when compression occurs.

Let's consider some stereo methods for specific uses.

STEREO TELEVISION

In stereo TV applications, stereo mic techniques are used mainly for Electronic News Gathering (E.N.G.), audience reaction, sports, and classical music.

☐ Imaging Considerations

Stereo TV presents some problems because of the disparity between picture and sound. Some viewers will be listening with speakers close to either side of the TV; others will be listening over a stereo system with speakers widely spaced. Where should the images be placed to satisfy both listeners? Should the sound-image locations coincide with the TV's visual images?

Extensive listening tests yielded these findings [1]:

- Listeners prefer dialog to be mono, or to have only a narrow stereo spread. A two-person interview with dialog switching from speaker to speaker is distracting.
- Listeners can accept off-camera sounds that originate away from the TV screen because the TV screen is considered a window on a larger scene.
- Listeners prefer sound images to be stable, even if the picture changes.
- Listeners easily notice a left-right reversal because the sound images don't match those on the screen.

Based on these results, many broadcasters recommend these practices for stereo TV:

1. Record dialog in mono in the center or with a narrow stereo spread.
2. Record effects, audience reaction, and music in stereo. You can allow off-camera sounds to be imaged away from the TV screen.
3. To prevent shifting images, don't move (pan) a stereo microphone once it is set up, and don't use camera-mounted stereo mics.
4. Avoid extreme differences between sound and picture. Be careful not to reverse left and right channels (say, by inverting an end-fired stereo microphone).

☐ Mono-Compatibility

An important requirement for any stereo broadcast is mono-compatibility. Since the majority of TV viewers are still listening in mono, the audio signal must sound tonally the same in mono or stereo. Also, excessive L−R signals in noncompatible programs can cause distortion and image shifting in AM stereo [2].

Phasing problems can arise when stereo signals with interchannel time differences are mixed to mono. Various frequencies are cancelled, causing a hollow or dull tone quality. Phase cancellations in mono also disrupt the relative loudness (balance) of instruments in a musical ensemble and change the volume of different notes played on the same instrument.

To ensure mono-compatibility, use a mono-compatible microphone array, such as a coincident pair, MS stereo microphone, or Crown SASS microphone. Also, when you use multiple close microphones panned for stereo, be sure to follow the 3:1 rule [3]. The distance between microphones should be at least three times the mic-to-source distance (as shown in Figure 8-1). This prevents phase cancellations if the stereo channels are summed to mono.

☐ Monitoring

Although headphones are often used for monitoring on-location, they can give a very different impression of stereo effects than loudspeakers.

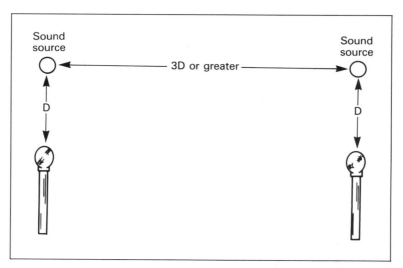

Figure 8-1 The 3:1 rule of microphone placement avoids phase interference between microphone signals.

For this reason, it's wise to monitor in a control room with small close-field monitor loudspeakers. These are placed about 3 ft from the listener to minimize the influence of room acoustics. Include a stereo/mono switch in the monitoring chain to listen for the problems described earlier [4].

How far apart should monitor speakers be for stereo TV listening? A typical TV viewer sees the TV screen covering a 16° angular width, while the optimum stereo listening angle for speakers is 60°. The disparity between these two angles can be distracting. A speaker angle of 30° (as shown in Figure 8-2) seems to be an adequate compromise for stereo TV listening [5, 6]. Keep the speakers at least 1 ft from the video monitor to prevent magnetic interference.

☐ Electronic News Gathering (E.N.G.)

The current practice in Electronic News Gathering is to pick up the correspondent's commentary in mono with a close-up handheld mic or lava-

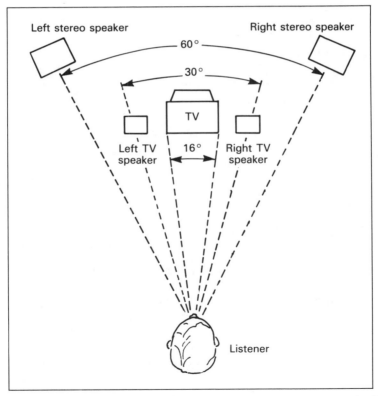

Figure 8-2 TV viewing angle (16°), stereo music listening angle (60°), and compromise stereo-TV listening angle (30°).

lier. This signal is panned to center between the loudspeakers. Stereo commentary is distracting: it's visually confusing to hear commentary to the left or right of the TV picture, removed from the image of the person speaking. Background ambience, however, is recorded in stereo for added realism. The mono commentary is usually mixed live with the stereo ambience.

Commentary and stereo ambience can also be mixed in post-production by taping the announcer and stereo background on a multitrack recorder [7]. If you use an M2 format VTR, however, this method can cause phasing problems. The M2 format recorder contains four audio tracks: two FM and two linear with Dolby C. The FM tracks are locked to speed, but the linear tracks are always shifting in time. Thus, the two stereo pairs of tracks continuously shift in phase relative to each other. For this reason, you may want to mix live if you use the M2 format.

A useful tool for on-location E.N.G. is a handheld stereo microphone. Currently, models of this type are made by Audio-Technica, AKG, Crown, Sony, and Sanken. The Sanken must be used with a shock-mount pistol grip. These microphones can be used just to pick up background sounds or can pick up both commentary and background with a single microphone. Figure 8-3 shows both types of E.N.G. pickup.

In the latter case, the reporter speaks directly in front of the microphone in the center. This results in a mono signal. The closer you place the mic to the reporter, the less background noise you pick up. But if background noise is excessive—for example, at a fire or a rock concert—the dialog may not be intelligible with this method.

☐ Audience Reaction

It's common in talk shows, game shows, or plays to pick up dialog in mono with close mics and cover audience reaction and music in stereo. Hearing the audience in stereo greatly enhances the feeling of being there. The close mics can either be standard mono units or MS Stereo units.

Stereo audience reaction can be picked up in many ways:

- with boundary mics on the walls, ceiling, or balcony edge
- with multiple mono microphones hung over the audience, panned for stereo
- with a single stereo microphone or mic pair
- with a combination of stereo mics and panned mono mics (Figure 8-4)

Spaced microphones give a desirable spacious feeling to audience reaction because of random phase relationships between channels.

Broadcast, Film and Video, Sound Effects, and Sampling □ 135

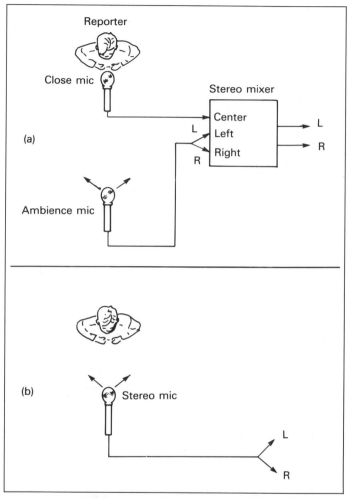

Figure 8-3 Typical stereo miking arrangements for E.N.G. (a) A spot mic for the reporter, plus a stereo mic for ambience. (b) A stereo mic picking up the reporter and ambience.

If the audience reaction is clearly imaged, it might be so realistic as to be distracting. For this reason, some engineers prefer to mix the audience mics to mono and run the mono mix through a stereo synthesizer.

If actors play near only one side of the audience, most of the reaction will be on that side. This one-sided reaction—left or right—may be confusing to home listeners. To hear this reaction on both sides in stereo, you may need to set up a stereo mic pair for each section of the audience.

136 ☐ STEREO MICROPHONE TECHNIQUES

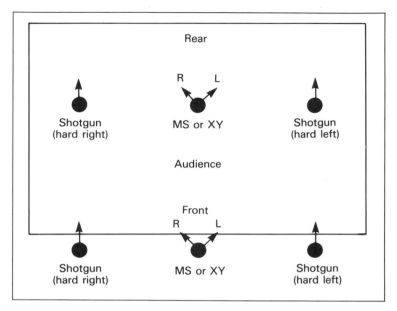

Figure 8-4 Stereo audience miking used for the Late Night with David Letterman and Pat Sajak shows.

The following tips on audience miking are from an article by Shawn Murphy [8]. The larger the audience, the fewer audience-reaction mics you need. Two or three mics can cover an audience of 3,000, while eight mics might be needed for an audience of 200.

It's important that the sound of the audience reaction match the sound of the dialog. This is more likely when you use the same type of microphone for both sound sources. When you turn dialog mics and audience mics on and off, the only change you should hear is an increase in the level of the P.A. system. Try to position the microphones to *not* pick up the sound-reinforcement speakers.

A typical signal-processing chain for the audience mics is as follows, in this order:

1. a stereo submixer to mix the mics to two channels
2. a foot pedal for volume control
3. a 100-Hz highpass filter to reduce room rumble

☐ **Parades**

Picking up parades clearly in stereo is difficult because of background noise. However, there are special stereo microphones that provide a

Broadcast, Film and Video, Sound Effects, and Sampling □ 137

tighter sound with less background pickup than standard stereo microphones. For example, Neumann offers an MS stereo shotgun microphone (see Chapter 9). The side element is a bidirectional capsule; the mid element is a short shotgun microphone. This microphone has been used to pick up parades in stereo. The Pillon Stereo PZM (Chapter 5) is another "tight sounding" microphone pair.

To pick up marching bands at the Mummer's Parade on New Year's Day in Philadelphia, KYW-TV3 used a pair of supercardioid boundary microphones placed back-to-back on a 2-ft-square piece of plexiglass. This arrangement was found to work better than conventional mics and shotguns. To reduce wind noise, the plexiglass boundary was wrapped in acoustically transparent fabric. This array was suspended on cables 32 ft over the street and 24 ft in front of the performance line of the bands (as shown in Figure 8-5).

To pick up the band as it approached the judges' stand, a similar array was mounted on a boom standing in the middle of the groups' performance area. Two backup mics were also mounted on the same boundaries [9].

Shotgun microphones are often needed in outdoor applications for isolation from the P.A. and background noise. Shotgun mics are preferred for sideline pickup of parades over parabolic microphones, because parabolics have a narrowband frequency response unsuitable for music.

Sound engineer Ron Streicher has used two shotguns crossed in an XY pattern for stereo pickup. He highpass-filters the shotguns to re-

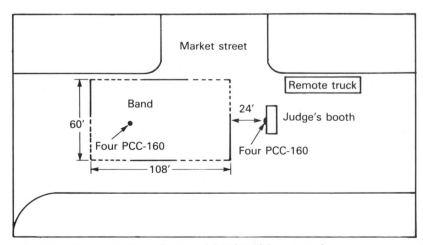

Figure 8-5 Stereo pickup of a marching band in a parade.

duce wind noise and mixes them with a single lowpass-filtered omni microphone to pick up low frequencies.

☐ Sports

Sporting events include three basic sound sources:

- the crowd reaction (including ambient acoustics)
- the sounds of the sport itself: players' yelling, baseball bat hits, tennis ball hits, bowling pin crashes, wrestling grunts, and so on
- the announcers' commentary

For crowd reaction and ambience, use an overall coincident pair or stereo microphone. If you use more than one such microphone, separate them widely. An alternative is to use widely spaced omnis or boundary mics. Widely spaced mics give poor image focus and a ping-pong effect but are more mono-compatible than closely spaced microphones.

For sport sounds, use close-up spot mics panned as desired. These mics are usually shotguns or parabolic reflectors. Mini cams can be equipped with stereo microphones, which are switched to follow the video shot.

For the announcers, use headset mics. A single announcer can be panned to center; two can be panned lightly left and right. Another method is to pan play-by-play to center; pan color #1 partly left and color #2 partly right. Roving reporters can use handheld wireless mics panned to center or panned opposite a booth announcer [10].

Sport sounds often require custom microphone pairs. Basketball games can be picked up with a stereo pair over center court, such as two boundary mics back-to-back on a panel, or you can use boundary mics on the floor at the edge of the court and on the backboards under the hoop. Indoor sports such as weight-lifting or fencing can be picked up with a near-coincident pair of directional boundary microphones or an MS boundary pair on the floor. For bowling, try a boundary mic on the back wall of the alley, high enough to avoid being hit, to pick up the pin action. Use a stereo pair on the alley itself [11].

A baseball game can be covered as suggested by Pizzi (Figure 8-6) [12]. Use a stereo mic near home plate, mixed with widely spaced boundary mics for ambience, plus shotgun or parabolic mics near the foul line aiming at the bases. Outfield microphones (cardioids or shotguns) are optional.

Multiple-sport events such as the Olympics can be covered with a number of MS mics or SASS mics at different locations. These mics pick up scattered events throughout the stadium or field.

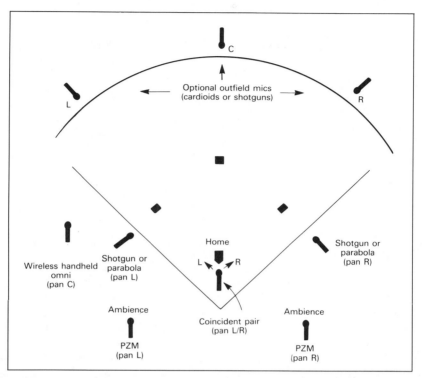

Figure 8-6 A suggested stereo microphone setup for baseball (after Pizzi).

■
STEREO RADIO

Stereo mic techniques are also needed for radio group discussions and plays.

□ Radio Group Discussions

For a group of people seated halfway around a table, try an XY pair (or a stereo microphone set to XY) over the center of the table (Figure 8-7). This will localize the images of the participants at various locations. To prevent phase cancellations due to sound reflections off the table, cover it with a thick pad or use an MS boundary microphone (described in Chapter 5)—or just remove the table.

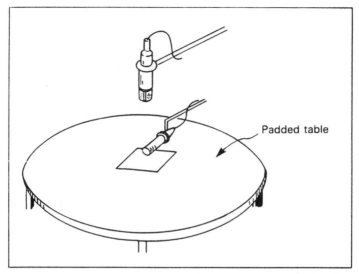

Figure 8-7 Picking up a radio group discussion with an XY stereo microphone over the center of the table or with an MS boundary microphone on the table.

When you broadcast a one-on-one interview, try to have the voices in mono and the ambience in stereo, because *ping-pong* interviews can be distracting. To do this, use an MS microphone and set the mid microphone to a bidirectional pattern. Place the participants on opposite sides of the mid microphone (front and rear). Control the amount of ambience by adjusting the side-microphone level [13]. An alternative is to use a Blumlein array (crossed figure eights) with the participants seated at the front and rear of the array.

☐ Radio Plays

Try a coincident pair, stereo microphone, or dummy head (for headphone reproduction) aiming at the center of the action. The actors can simulate distance by moving away from the microphones. Good-sounding positions can be marked with tape on the floor.

It's important that the actors don't go beyond the pickup angle of the stereo mic pair. This is especially true for the Blumlein pair of crossed figure eights, because the left and right channels will be in opposite polarity if the actor is past the axis of either microphone.

To determine the stereo pair's pickup angle, monitor the output of the pair in stereo while someone speaks in front of the pair. Have that person start in the center and slowly move to one side while speaking. When the monitored image of the person speaking is all the way to one speaker, have the person stop and mark the floor with tape at that point. Do the same in the other direction off-center. Tell the actors to stay within the two tape marks.

You may want to supply the actors with stereo headphones so that they can hear the effects of their position and closeness to the mics. However, note that headphone monitoring of a coincident pair has little stereo separation.

FILM AND VIDEO

Stereo miking also finds use in feature films, as well as video documentaries and industrial films.

☐ Feature Films

Sound engineers for feature films usually pick up dialog with one or two shotgun mics or lavalier mics to reduce ambience. During post production, the mono signals from these microphones are panned to the desired stereo locations to match the visual locations of the actors on- or off-screen. This forms a dialog premix. Often the original dialog track recorded on-location is unusable, so it is replaced in the looping process.

The final sound track is a stereo mix of several stereo premixes:

1. dialog premix
2. sound effects premix
3. ambience or atmosphere premix
4. music score premix

An alternative to panning is to record dialog in stereo with a stereo shotgun mic on a boom, with a PZM stereo shotgun, or with a SASS microphone. The stereo perspective of the sound should match the visual perspective on a theatre screen.

You might want to add a stereo mic or stereo pair for ambience. The realism of background ambience or *wild sound* is enhanced by stereo recording.

Sennheiser has suggested a novel MS stereo miking method for films, TV, and so on: the side element of the MS array is a stationary side-aiming bidirectional microphone; the mid element is a boom-mounted shotgun mic which you aim at the sound source. The shotgun is physically panned to follow the action [14].

☐ Documentaries and Industrial Productions

For these productions, you can either

- pick up both dialog and ambience with a single stereo microphone (handheld or stand-mounted), or
- use a closeup mic for the announcer, mixed with a stereo array for ambience (as shown in Figure 8-3).

As for positioning the stereo microphone, hold it up high at arm's length above the camera, aiming down at the sound source, or hold the mic below the camera, aiming up at the sound source. A stereo microphone can be attached to a Steadicam platform to follow the action. The first use of this method was by Gary Pillon of General Television Network, Oak Park, Michigan. Using a stereo PZM microphone that he had designed, he recorded the sound track for a documentary that subsequently won an Emmy.

Another way to pick up stereo ambience is to use a coincident pair of short shotgun microphones. They produce a wide stereo effect with a hole-in-the-middle. This hole can be filled with the announcer's voice in mono [15].

For outdoor ambience, you might try a pair of omni microphones spaced more than 25 ft apart. Since the coherence between microphones is small at that spacing, mono cancellation is reduced. Also, the omni mics are relatively insensitive to wind noise [16].

SOUND EFFECTS

Recording sound effects in stereo requires some special considerations. When you record moving sound sources, such as truck or plane passes, you need a stereo mic array that accurately tracks the motion. Coincident and near-coincident techniques work well in this application. With spaced-microphone recordings, the sound image usually jumps from one speaker to the other as the sound source passes the mic array. Accurate motion tracking in sound-effect recording is analogous to accurate localization in music recording.

When recording a moving effect, experiment with the distance between the microphone and the closest pass of the sound source. The closer the microphone is to the path of the subject, the more rapidly the image will pass the center point (almost hopping from one channel to the other). To achieve a smooth side-to-side movement, you may need to increase the distance [17].

Since you'll be monitoring your recording in the field with headphones, you may want to use a stereo miking technique that provides the same stereo spread on headphones as on loudspeakers. Recommended for this purpose are near-coincident arrays, such as ORTF, NOS, OSS, or the SASS microphone.

For recording animals, some recordists hide the microphones in shrubbery near the animals and run long cables back to the recorder so they won't inhibit the animals.

An award-winning nature recordist, Dan Gibson has recorded soundtracks for wildlife films and for the Solitudes series of records. He records digitally with a Sony DAT recorder. For birds and long-range pickup, Gibson uses a stereo parabolic microphone P200S, for mid-range distances he uses a pair of Neumann shotgun mics, and for close-up work he uses a pair of AKG D 222 microphones.

To edit out noises in his DAT recordings, he first makes a copy from one DAT to another. He plays the first recording up to the noise, then quickly crossfades to the other recording cued up just after the noise.

Some miscellaneous tips on sound-effect recording:

- Use a shock mount and windscreen outdoors.
- If you want to reduce pickup of local ambience and background noises, place the mic close to the sound source.
- Record flat; most effects need EQ later.
- Record multiple takes so that you can choose the best one.

SAMPLING

Sampling is the process of digitally recording a short sound event, such as a single note from a musical instrument. The recording is stored in computer memory chips. You play it back by pressing keys on a piano-style keyboard. The higher the key, the higher the pitch of the reproduced sample.

Stereo sampling is a new feature currently available in Synclavier and E-mu systems. If you play, for instance, a stereo piano sample, each note is reproduced in its original spatial location.

When sampling, be sure to use microphones with low self-noise (for example, less than 20 dBA). If the sample is noisy, the noise will audibly shift in pitch as you play different keys on your sampling keyboard. Variable-pitch noise is easier to hear than steady noise.

Another requirement for a stereo sampling microphone is a wide, flat frequency response. This avoids tonal coloration of the sample. A hypercardioid or supercardioid pattern, as well as close microphone placement, can help to reduce pickup of ambient noise and room reflections, so that you hear only the desired sound source.

When you record samples or sound effects for keyboard, drum-machine, or disk-soundbank reproduction, any recorded ambience will be reproduced as part of the sample. For added future flexibility, you may want to make several samples of one source at different distances to include the range of added reverberance.

Off-center images can be reproduced accurately by sampling the sound source in the desired angular position as perceived from stereo center. Recorded ambience will sharpen the image, but is not necessary.

When you pitch-shift a stereo sample, the image location and the size of the room will change if you use spaced or near-coincident techniques. That's because the interchannel delay varies with the pitch shift. To minimize these undesirable effects, try sampling at intervals of one-third octave or less, or record the sample with a coincident-pair technique [18].

When looping, try to control the room ambience so that it is consistent before and after the sample (unless reverberant decay is desired as part of the sample).

REFERENCES

1. F. Rumsey, *Stereo Sound for Television*, Boston: Focal Press, 1989, pp. 13, 14, 41.
2. S. Pizzi, "Stereo Microphone Techniques for Broadcast," *Audio Engineering Society Preprint* No. 2146 (D-3), presented at the 76th Convention, October 8–11, 1984, New York.
3. Ibid.
4. Ibid.
5. "The Link," *High Fidelity*, 1984 (Nov.).
6. P. Lehrman, "Multichannel Audio Production for US TV," *Studio Sound*, 1984 (Sept.), pp. 34–36.
7. S. Pizzi, "A 'Split Track' Recording Technique for Improved E.N.G. Audio," *Audio Engineering Society Preprint* No. 2016, presented at the 74th Convention, October 1983.

8. S. Murphy, "Live Stereo Audio Production Techniques for Broadcast Television," *The Proceedings of the AES 4th International Conference on Stereo Audio Technology for Television and Video*, Rosemont, IL, May 5–18, 1986, pp. 175–79.

9. *Mic Memo*, April 1988. Crown International, 1718 W. Mishawaka Rd., Elkhart, IN 46517.

10. Pizzi, "Stereo."

11. Crown Microphone Catalog, Crown International, 1718 W. Mishawaka Rd., Elkhart, IN 46517.

12. Pizzi, "Stereo."

13. Ibid.

14. H. Gerlach, "Stereo Sound Recording With Shotgun Microphones," *Journal of the Audio Engineering Society*, Vol. 37, No. 10, 1989 (October), pp. 832–838.

15. Pizzi, "Stereo."

16. Ibid.

17. Crown SASS Owner's Manual (material for this portion was provided by Michael Billingsley).

18. Ibid.

A special thanks to Terry Skelton, staff audio instructor at NBC, for his helpful suggestions on this chapter.

9

Stereo Microphones and Accessories

☐

This is a listing of stereo microphones, dummy heads, MS matrix boxes, and stereo microphone stand adapters. The list is up to date only for the date of publication of this book. Since models and prices will change, please contact the manufacturers for current information. This chapter should be viewed less as a catalog and more as an illustration of what features are currently available. Prices range from $500 to $3500 (except for the Calrec Soundfield Microphone and Radio Shack microphones). Regarding stereo microphone specifications, please note the following definitions:

Side-fired: The axis of maximum sensitivity is at right angles to the microphone's longitudinal axis; you hold the mic vertically.

End-fired: The axis of maximum sensitivity is the same as the microphone's longitudinal axis; you point the mic at the center of the sound source.

■ STEREO MICROPHONES

AKG C-34 small-diaphragm condenser microphone for MS or XY recording: nine patterns remotely selectable; one capsule is fixed, the other is rotatable through 180°; side-fired

AKG C-422 large-diaphragm condenser microphone for MS or XY recording: nine patterns remotely selectable; one capsule is fixed, the other is rotatable through 180°; side-fired (see Figure 9-1)

AKG C-522 ENG: two fixed cardioid condenser capsules for XY recording of news, sports, etc.; end-fired (see Figure 9-2)

Stereo Microphones and Accessories □ 147

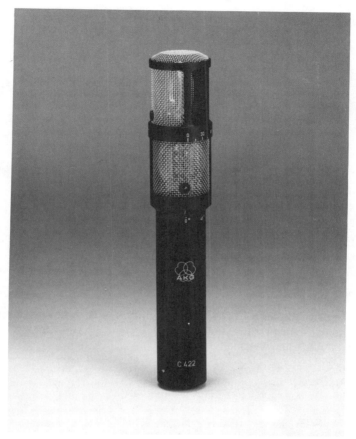

Figure 9-1 AKG-C-422 stereo microphone. (Courtesy AKG Acoustics, Inc.)

AKG C-426B "Comb": stereo microphone with two twin-diaphragm condenser capsules, one atop the other, for MS or XY use; has three patterns plus six intermediate steps; includes low-cut switch, attenuator, and shock mount; side-fired

AMS/Calrec Soundfield Microphone: uses four capsules arranged in a tetrahedron, phase-matrixed for true coincidence; remote control of polar pattern, azimuth (horizontal rotation), elevation (vertical tilt), and dominance (fore/aft movement); separate outputs for stereo and Ambisonic (3-D) surround sound; side-fired; with flight case and 20 meter head cable

AMS/Calrec ST250 coincident-stereo condenser microphone with MS or L/R outputs, variable polar patterns, attenuator, angle control: includes

148 □ STEREO MICROPHONE TECHNIQUES

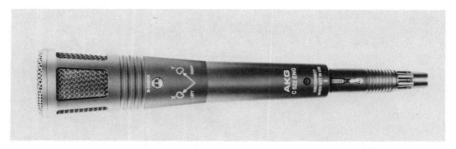

Figure 9-2 AKG-C-522 ENG stereo microphone. (Courtesy AKG Acoustics, Inc.)

separate stereo control unit operated on internal C cells or phantom powered; end-fired or side-fired

Audio-Technica AT-825 XY stereo microphone: low-cut and battery/phantom powering; end-fired

Beyerdynamic MC742: XY or MS microphone with rotatable upper capsule and remote-adjustable polar patterns; 10 dB pad and low-cut filter; side-fired

Crown SASS-P PZM Stereo Microphone: two ear-spaced PZMs on angled boundaries; for near-coincident, mono-compatible recording, ENG, etc.; end-fired

Crown SASS-B: shaped like the SASS-P, a stereo boundary mount for Bruel & Kjaer 4006 low-noise studio microphones

Fostex M22RP Mid-Side microphone with printed ribbon diaphragms: no power supply needed; three outputs: cardioid mid, cardioid right side, cardioid left side; transformer box included with left and right outputs; most suited for outdoor uses, sports broadcasts, and auditorium on-air monitoring; side-fired

Fostex M20RP Mid-Side microphone: low-cost version of the M22RP; has 3-way cable rather than matrix; side-fired

Josephson Engineering OSS-Disk (Jecklin Disk): uses two omnidirectional pressure-response microphones spaced $6\frac{1}{2}$ inches and separated by an acoustically damped disk $11\frac{7}{8}$ inches in diameter; uses time and spectral differences to localize stereo images; end-fired

Josephson Engineering C-622 OSS-II Stereo Boundary Condenser Microphone: boundary version of the OSS disk, uses two boundary microphones spaced $6\frac{1}{2}$ inches and separated by a half-circle plastic disk covered with absorbent material; uses time and spectral differences to localize sound images

Josephson Engineering SCH-Disk (Schneider Disk): new version of OSS disk with improved phase linearity and imaging due to hemispheric foam baffles between mics and disk

Josephson Engineering Stereo-Q: four cardioid capsules are mounted in an X, and the four discrete signals are output to a stereo matrix unit to produce left and right output signals; angle of X may be set to detented positions of 60°, 90°, or 120° for narrow, normal, or wide XY pickup; provides MS and crossed figure-eight pickup; side-fired

Neumann SM69 fet: multi-pattern remote-controlled dual stereo MS/XY microphone; upper capsule can be rotated relative to the lower one through 270°; side-fired (see Figure 9-3)

Neumann USM 69i: same as Neumann USM SM69 fet, but with directional pattern switches located on the microphone itself, eliminating the need for the in-line box; side-fired

Neumann RSM 190S System: MS stereo shotgun microphone, uses a short shotgun for the mid element and a side-facing bidirectional capsule for the side element; M and S outputs feed a MTX 190 S matrix amplifier for remote control of stereo spread (mid/side ratio); end-fired (see Figure 9-4)

Neumann RSM 191S: recent upgrade of RSM 190S

Pearl TL-4: two back-to-back cardioids for XY recording; end- fired

Pearl MS-2: mid-side microphone with internal MS matrix; has left and right outputs; end-fired

Pearl MS-8: mid-side microphone without internal MS matrix; has mid and side outputs; end-fired

Figure 9-3 Neumann SM69 fet stereo microphone. (Courtesy Gotham Audio Corporation.)

150 □ STEREO MICROPHONE TECHNIQUES

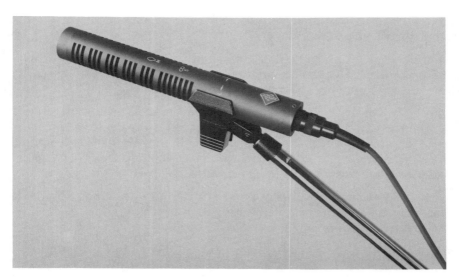

Figure 9-4 Neumann RSM 190S stereo shotgun microphone. (Courtesy Gotham Audio Corporation.)

Radio Shack (Realistic) 33-2012: XY stereo microphone, battery powered, 600-ohm unbalanced outputs, switch-selectable normal or wide stereo spread; end-fired

Radio Shack (Realistic) 33-1065: near-coincident stereo microphone with adjustable angling; battery powered, 600-ohm unbalanced outputs; end-fired

Sanken CMS-2 MS Stereo Condenser Microphone: small, lightweight, handheld unit for TV, film, and music; cardioid mid element; bidirectional side element; end-fired (see Figure 9-5)

Sanken CMS-7 MS Stereo Portable Condenser Microphone: for broadcast and film; requires external matrix box; fixed angle of polar patterns; Model CMS-7 has cardioid mid unit; CMS-7H has hypercardioid mid unit; end-fired (see Figure 9-6)

Sanken CMS-9 MS Stereo Portable Condenser Microphone: similar to CMS-7, but with internal matrix so no external box required; left/right or MS outputs, fixed angle of polar patterns; end-fired

Schoeps MSTC 34, 44, and 54 stereo twin microphone: two cardioid condenser capsules permanently angled 110° apart and spaced 17 cm (ORTF system); end-fired

Schoeps CMTS 501U: intensity stereo microphone for MS or XY recording; upper capsule can be rotated over 360° and is used for channel 2;

Figure 9-5 Sanken CMS-2 stereo microphone. (Courtesy Pan Communications, Inc.)

fixed capsule is for channel 1; each capsule's pattern is switch-selectable at the microphone to omni, cardioid, or bidirectional; side-fired (see Figure 9-7)

Shure VP88: MS microphone with stereo or MS outputs, switchable stereo spread control on microphone, low-cut switch; end-fired

Sony ECM-979: MS electret condenser microphone with internally adjustable stereo spread; 1.5 V battery power only; left and right outputs with dual XLR-type cable; side-fired

Sony ECM-MS5: MS electret condenser microphone, light weight for field use; has internally adjustable stereo spread, low-cut switch, phantom or AA battery power plus optional DC power supply; left and right outputs with dual XLR-type cable; end-fired

152 □ STEREO MICROPHONE TECHNIQUES

Figure 9-6 Sanken CMS-7 stereo microphone. (Courtesy Pan Communications, Inc.)

Figure 9-7 Schoeps CMTS 501U stereo microphone. (Courtesy Posthorn Recordings.)

DUMMY HEADS

Aachen Head model HMS II: dummy head with measurement microphones; record processor equalizes the head signals to have flat response in a frontal free-field; reproduce unit contains free-field equalizers for use with Stax SR-Lambda Professional™ headphones; also available for recording is a unit with Schoeps capsules, two XLR-type outputs, and switchable EQ for frontal free-field or independent-of-direction field

Bruel & Kjaer 4128 Head and Torso Simulator: uses two B&K 4006 microphones; special-order item

Coles Biophonic Mr. Aural and Lady Aural heads: based on accurate skin softness and internal density

Holophonic Systems H-1 Holophone: "Based on the exact reproduction of the human head, with specially designed acoustic membranes and a semiconductor signal processor"; self-contained windscreen; available by rental only

Neumann KU 81i: dummy head binaural system, including simplified "Fritz II" dummy head with internal microphones, AC or battery supply, and cables; stereo compatible; acoustically equalized internally for flat diffuse field response; auditory canal shortened to 4 mm (the shortest length needed for directional effects) (see Figure 9-8)

Sennheiser MKE 2002 Set: pair of miniature omni condenser microphones mounted on a headset, worn in your ears or placed in the ears of a simplified dummy head; dummy head with no ear canals included; not equalized for head diffraction

The KEMAR dummy head by Knowles Electronics: designed for research rather than for dummy-head recording

STEREO MICROPHONE ADAPTERS (STEREO BARS)

AKG KM-235/1: V-shaped; adjustable stereo bar or rail

AKG H 10: metal stereo crossbar with two $\frac{3}{8}$ inch knurled head screws; can be spaced $1\frac{5}{8}$ inches to 3 inches

154 □ STEREO MICROPHONE TECHNIQUES

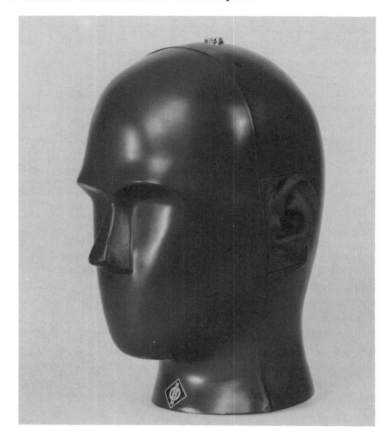

Figure 9-8 Neumann KU 81i dummy head. (Courtesy Gotham Audio Corporation.)

AKG H 52: stereo suspension set for condenser microphone capsules CK 1 X or CK 3 X; for mounting on mic stand or hanging from ceiling; for coincident or near-coincident arrays

Beyer ZMS1: V-shaped; adjustable mounting rail; fits two microphones for stereo recordings with a maximum separation of 8 inches

Bruel & Kjaer 3529 and 3530 stereo microphone sets: adjustable linear stereo rail; includes microphones and power supplies

Neumann DS21 dual microphone mount: can be used to combine two miniature microphones and two bent capsule extension tubes into one fixed assembly for stereo recordings

Sanken XY stereo microphone mount: used to set a pair of CU-41 microphones with S-41 adapters for XY stereo recording

Stereo Microphones and Accessories □ 155

Figure 9-9 Schoeps UMS 20 Universal stereo bracket. (Courtesy Posthorn Recordings.)

Schoeps UMS 20 Universal stereo bracket: allows horizontal or vertical mounting of a coincident or near-coincident pair; fully adjustable (see Figure 9-9)

Schoeps UMSC Universal Colette stereo bracket: positions two capsules with KC-Collette cables for coincident or near-coincident recording

Schoeps STC ORTF system stereo mounting bar: positions two cardioid capsules with KC cables for ORTF stereo

Sennheiser MZT 235 and MZS 235 adapter bars: fixed spacing, adjustable angle

Shure Model A27M: uses two rotating, stacked cylinders; allows angling and spacing adjustment for coincident and near-coincident methods (see Figure 9-10)

MATRIX DECODERS

Audio Engineering Associates MS38 dual-mode active MS matrix: input switchable MS or left/right; adjustable mid/side ratio (stereo spread control)

156 □ STEREO MICROPHONE TECHNIQUES

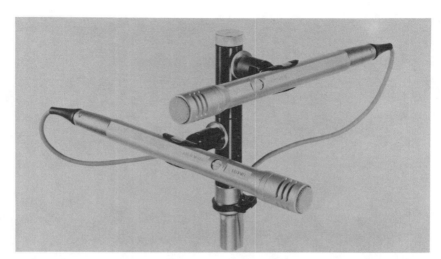

Figure 9-10   Shure A27M stereo microphone adapter. (Courtesy Shure Brothers, Inc.)

Audio Engineering Associates MS380 TX: similar to MS38 plus high-pass filter, pad, gain trims, mic pre/matrix switch, and master level (see Figure 9-11)

Neumann Z 240: matrixing transformer pair

Sanken MBB-II battery power supply and switchable matrix box for the Sanken CMS-7 microphone: MS or left/right output switch (see Figure 9-12)

Sanken M-5 Matrix Box for the Sanken CMS-2 microphone: dual transformers

Figure 9-11   Audio Engineering Associates MS380 TX Universal Matrix Box. (Courtesy Audio Engineering Associates.)

Stereo Microphones and Accessories □ **157**

Figure 9-12   Sanken MBB-II matrix box. (Courtesy Pan Communications Inc.)

Schoeps VMS 52 UB, 32 UB microphone preamp, MS matrix and power supply: low-cut filter, gain and width controls; powered by batteries or an external power supply

Gotham Audio offers a PC software program written by John Woram, *The MS Polar Pattern Program*. The user enters distance from the stage, stage width, and MS polar patterns. The program plots the mid pattern, side pattern, sum and difference, and mono and stereo angles.

COMPANY ADDRESSES

Aachen Head
c/o Electronic Architecture Techniques, Inc.
114A Washington St.
Norwalk, CT 06854
Tel. (203) 838-4167

Main office:

Dr. Klaus Genuit, Head Acoustics GmbH
Kaiserstr. 100
D-5120 Herzogenrath 3, Germany
Tel. (0) 2407-577-0

AKG Acoustics, Inc.
77 Selleck St.
Stamford, CT 06902
Tel. (203) 348-2121

Audio Engineering Associates
1029 N. Allen Ave.
Pasadena, CA 91104
Tel. (818) 798-9127

Telescoping mic stands also available.

Audio Technica U.S., Inc.
1221 Commerce Drive
Stow, OH 44224
Tel. (216) 686-2600

Beyerdynamic, Inc.
5-05 Burns Ave.
Hicksville, NY 11801
Tel. (516) 935-8000

Bruel & Kjaer Instruments, Inc.
185 Forest St.
Marlborough, MA 01752
Tel. (617) 481-7000

Calrec Audio Ltd.
AMS Industries Park
Burnley, Lancashire BB11 5ES
England
Tel. (0282) 57011

Also available at:

AMS Industries Inc.
P.O. Box 31864
3827 Stone Way North
Seattle, WA 98103
Tel. (206) 633-1956

Crown International
1718 W. Mishawaka Rd.
Elkhart, IN 46517
Tel. (219) 294-8000

Fostex Corporation of America
15431 Blackburn Ave.
Norwalk, CA 90650
Tel. (213) 921-1112

Gotham Audio Corp.
1790 Broadway
New York, NY 10019-1412
Tel. (212) 765-3410

Stereo Microphones and Accessories □ 159

Holophonic Systems S.P.A.
Via Magno Magnini, 24-06100
Perugia, Italy
Tel. (075) 751715-74770-756522

Sales offices:

Viale Parioli
101/C-00197
Roma, Italy
Tel. (06) 802990-874354

Josephson Engineering
3729 Corkerhill
San Jose, CA 95121
Tel. (408) 238-6062

Neumann Microphones
c/o Georg Neumann GMBH
Charlottenstrasse 3
D-1000 Berlin 61
Tel. (030) 2 59 93-0

Also:

c/o Gotham Audio Corporation
1790 Broadway
New York, NY 10019-1412
Tel. (212) 765-3410

Pearl Mikrofonlab
P.O. Box 98 265 01
Astorp, Sweden
Tel. 042-588 10 +046-42 588 10

Also available at:

Camera Mart
456 W. 55th St.
New York, NY 10019
Tel. (212) 757-6977

Sanken Microphones, Pan Communications Inc.
Suite 607, 1-5-10 Roppongi, Minato-ku
Tokyo 106, Japan
Tel. +81 (0) 3 505 5463

Also:

Audio Intervisual Design
1032 N. Sycamore
Los Angeles, CA 90038
Tel. (213) 469-4773

Schoeps Microphones
c/o Posthorn Recordings
142 W. 26th St.
New York, NY 10001
Tel. (212) 242-3737

Sennheiser Electronic Corporation
6 Vista Drive
P.O. Box 987
Old Lyme, CT 06371
Tel. (203) 434-9190

Shure Brothers Inc.
222 Hartrey Ave.
Evanston, IL 60204
Tel. (312) 866-2200

Sony Communications Products Co.
Sony Drive
Park Ridge, NJ 07656
Tel. (201) 939-1000

GLOSSARY

A-B: See Spaced pair.

Accent microphone: See Spot microphone.

Ambience: Room acoustics; early reflections and reverberation; also, the audible sense of a room or environment surrounding a recorded instrument.

Ambience microphone: A microphone placed relatively far from its sound source to pick up ambience.

Amplitude: Level, intensity, or magnitude. The amplitude of a sound wave or signal, as measured on a meter, is 0.707 times the peak amplitude. The peak amplitude is the voltage of the signal waveform peak.

Artificial head: See Dummy head.

Balance: The relative volume levels of various instruments in a musical ensemble.

Bidirectional microphone: A microphone that is most sensitive to sounds arriving from two directions—in front of and behind the microphone. It rejects sounds approaching from either side of the microphone. Also called a cosine or figure-eight microphone because of the shape of its polar pattern.

Binaural recording: A 2-channel recording made with an omnidirectional microphone in each ear of a human or a dummy head, for playback over headphones. The object is to duplicate the acoustic signal appearing at each ear.

Blumlein array: A stereo microphone technique in which two coincident bidirectional microphones are angled 90° apart (45° to the left and right of center).

Boundary microphone: A microphone designed to be used on a boundary (a hard reflective surface). The microphone capsule is mounted very close to the boundary so that direct and reflected sounds arrive at the microphone diaphragm in phase (or nearly so) at all frequencies in the audible band.

Capacitor microphone: See Condenser microphone.

Cardioid microphone: A unidirectional microphone with side attenuation of 6 dB and maximum rejection of sound at the rear of the microphone (180° off axis). A microphone with a heart-shaped directional pattern.

Channel: A single path of an audio signal. Usually, each channel contains a different signal.

Closely spaced: See Near-coincident.

Coincident pair: A stereo microphone, or two separate microphones placed so that the microphone diaphragms occupy approximately the same point in space, angled apart and mounted one directly above the other. Intensity or amplitude differences between channels are used to localize sound images.

Comb-filter effect: The frequency response caused by combining a sound with its delayed replica. The frequency response has a series of peaks and dips caused by phase interference. The peaks and dips resemble the teeth of a comb. This effect can occur with near-coincident and spaced-pair techniques when the left-and-right-channel signals are combined to mono.

Compressor: A signal processor that reduces dynamic range by means of automatic volume control; an amplifier whose gain decreases as the input signal level increases above a preset point.

Condenser microphone: A microphone that works on the principle of variable capacitance to generate an electrical signal. The microphone diaphragm and an adjacent metallic disk (called a backplate) are charged with static electricity to form two plates of a capacitor. Incoming sound waves vibrate the diaphragm, varying its spacing to the backplate, which varies the capacitance, which varies the voltage between the diaphragm and backplate.

Crosstalk: The unwanted transfer of a signal from one channel to another. Head-related crosstalk is the right-speaker signal that reaches the left ear and the left-speaker signal that reaches the right ear. In the transaural stereo system, this acoustic crosstalk is cancelled by processing the stereo signal with electronic crosstalk that is the inverse of the acoustic crosstalk.

dB: Abbreviation for decibel.

Dead: Having very little or no reverberation.

Decibel: The unit of measurement of audio level; ten times the logarithm of the ratio of two power levels; twenty times the logarithm of the ratio of two voltages.

dBV is decibels relative to 1 volt.

dBu is decibels relative to .775 volt.

dBm is decibels relative to 1 milliwatt.

dBA is decibels, A-weighted (see Weighted)

Delay: The time interval between a signal and its repetition. A digital delay or a delay line is a signal processor that delays a signal for a short time. Interchannel delay can occur acoustically between spaced microphones.

Depth: The audible sense of nearness and farness of various instruments. Instruments recorded with a high ratio of direct-to-reverberant sound are perceived as being close; instruments recorded with a low ratio of direct-to-reverberant sound are perceived as being distant.

Diffuse field: A sound field in which the sounds arrive randomly from all directions, such as the reverberant field in a concert hall. Diffuse-field equalization is equalization applied to a dummy head so that it has a net flat response in a diffuse sound field.

Directional microphone: A microphone that has different sensitivity in different directions; a unidirectional or bidirectional microphone.

Direct sound: Sound traveling directly from the sound source to the microphone (or to the listener) without reflections.

Distortion: An unwanted change in the audio waveform, causing a raspy or gritty sound quality; the appearance of frequencies in a device's output signal that were not in the input signal.

Dry: Having no echo or reverberation; close-sounding.

Dummy head: A modeled head with microphones in the ears, used for binaural recording; same as artificial head.

Dynamic microphone: A microphone that generates electricity when sound waves cause a conductor to vibrate in a stationary magnetic field. The two types of dynamic microphone are moving coil and ribbon. A moving-coil microphone is usually called a dynamic microphone.

Dynamic range: The range of volume levels in a program from softest to loudest.

Echo: A delayed repetition of a signal or sound; a sound delayed 50 milliseconds or more, combined with the original sound.

Electret-condenser microphone: A condenser microphone in which the electrostatic field of the capacitor is generated by an electret: a material that permanently stores an electrostatic charge.

Electrostatic field: The force field between two conductors charged with static electricity.

Elevation: An image displacement in height above the speaker plane.

Envelope: The rise and fall in volume of one note. The envelope connects successive peaks of the waves comprising a note. Each harmonic in the note might have a different envelope.

Equalization (EQ): The adjustment of frequency response to alter the tonal balance or to attenuate unwanted frequencies.

Equalizer: A circuit (usually in each input module of a mixing console, or in a separate unit) that alters the frequency spectrum of a signal passed through it.

Faulkner method: Named after Tony Faulkner, a stereo microphone technique using two bidirectional microphones aiming at the sound source and spaced about 8 inches apart.

Focus: The degree of fusion, compactness, or positional definition of a sonic image.

Free field: The sound field coming directly from the sound source without reflections; the sound field in an anechoic chamber. Free-field equalization is equalization applied to a dummy head to make it have a net flat response in a free field.

Frequency: The number of cycles per second of a sound wave or an audio signal, measured in hertz (Hz). A low frequency (for example, 100 Hz) has a low pitch; a high frequency (for example, 10,000 Hz) has a high pitch.

Frequency response: The range of frequencies that an audio device will reproduce at an equal level (within a tolerance, such as ±3dB).

Fundamental: The lowest frequency in a complex wave.

Fusion: The formation of a single image by two or more sound sources, such as loudspeakers.

Generation: A copy of a tape. A copy of the original master recording is a first generation tape. A copy made from the first generation tape is a second generation, and so on.

Generation loss: The degradation of signal quality (the increase in noise and distortion) that occurs with each successive generation of a tape recording.

Harmonic: In a complex wave, an overtone whose frequency is a whole-number multiple of the fundamental frequency.

Headphones: A head-worn transducer that covers the ears and converts electrical audio signals into sound waves.

Hertz (Hz): Cycles per second, the unit of measurement of frequency.

Hiss: A noise signal containing all frequencies, but with greater energy at higher octaves. Hiss sounds like wind blowing through trees. It is usually caused by random signals generated by microphones, electronics, and magnetic tape.

Hypercardioid microphone: A directional microphone with a polar pattern that has 12 dB attenuation at the sides, 6 dB attenuation at the rear, and two nulls of maximum rejection at 110° either side off axis.

Glossary □ **165**

Image: An illusory sound source reproduced by two or more speakers, located somewhere relative to the speakers (usually between them).

Imaging: The ability of a microphone array or speaker pair to form easily localizable images.

Input module: In a mixing console, the set of controls affecting a single input signal. An input module usually includes an attenuator, fader, equalizer, effects send, cue send, and channel-assign controls.

Intensity stereo (XY stereo): A method of forming stereo images by intensity or amplitude differences between channels; see Coincident pair.

ITE/PAR: Acronym for In The Ear/Pinna Acoustic Response, a stereo recording system developed by Don and Carolyn Davis of Synergetic Audio Concepts. It uses two probe microphones in the ear canals, near the ear drum of a human listener. Playback is over two speakers up front and two to the sides of the listener.

Jecklin disk: Named after its inventor, a stereo microphone array using two omnidirectional microphones spaced $6\frac{1}{2}$ inches apart and separated by a disk or baffle $11\frac{7}{8}$ inches in diameter, covered with sound-absorbent material; also known as the OSS system.

Kilo: A prefix meaning one thousand; abbreviated k.

Level: The degree of intensity of an audio signal: the voltage, power, or sound pressure level. The original definition of level is the power in watts.

Live: Having audible reverberation.

Live recording: A recording made at a concert; a recording made of a musical ensemble playing all at once, rather than overdubbing.

Localization: The ability of the human hearing system to tell the direction of a real or illusory sound source; the relation between interchannel or interaural differences and perceived image location.

Localization accuracy: The accuracy with which a stereo microphone array translates the location of real sound sources into image locations. If localization is accurate, instruments at the side of the musical ensemble are reproduced from the left or right speaker; instruments halfway off-center are reproduced halfway between the center of the speaker pair and one speaker, and so on.

Location: The angular position of an image relative to a point straight ahead of a listener, or its position relative to the loudspeakers.

Loudspeaker: A transducer that converts electrical energy (the signal) into acoustical energy (sound waves).

Mic: An abbreviation for microphone.

Mic level: The level or voltage of a signal produced by a microphone, typically 2 millivolts.

Microphone: A transducer or device that converts an acoustical signal (sound) into a corresponding electrical signal.

Microphone techniques: The selection and placement of microphones to pick up sound sources.

Mid-side: A coincident-pair stereo microphone technique using a forward-facing unidirectional, omnidirectional, or bidirectional microphone and a side-facing bidirectional microphone. The microphone signals are summed and differenced to produce left- and right-channel signals.

Mike: To pick up with a microphone.

Mix: To combine two or more different signals into a common signal.

Mixer: A device that mixes or combines audio signals and controls the relative levels of the signals.

Monitor: To listen to an audio signal with headphones or a loudspeaker. A monitor is a loudspeaker in a control room used for monitoring.

Monaural: Referring to listening with one ear; often incorrectly used to mean monophonic.

Mono, monophonic: Referring to a single channel of audio. A monophonic program can be played over one or more loudspeakers, or one or more headphones.

Mono-compatible: A characteristic of a stereo program, in which the program channels can be combined to a mono program without altering the frequency response or balance. A mono-compatible stereo program has the same frequency response in stereo or mono because there is no delay or phase shift between channels to cause phase interference.

MS recording: See Mid-side.

Muddy: Unclear sounding; having excessive reverberation.

Near-coincident: A stereo microphone technique in which the two microphones are angled apart symmetrically on either side of center and horizontally spaced a few inches apart. Both time and amplitude differences between channels localize the sound images.

Near-field™ monitoring: A monitor-speaker arrangement in which the speakers are placed very near the listener (usually on top of the mixing console meters), to reduce the audibility of control-room acoustics and improve stereo imaging (trademark of Ed Long Associates).

Noise: Unwanted sound, such as hiss from electronics or tape; an audio signal with an irregular, nonperiodic waveform.

N.O.S.: A near-coincident stereo microphone technique in which two cardioid microphones are angled apart 90° and spaced 30 cm horizontally.

Off axis: Not directly in front of a microphone or loudspeaker.

Off-axis coloration: In a microphone, the deviation from the on-axis frequency response that sometimes occurs at angles off the axis of the microphone. The coloration of sound (alteration of tone quality) for sounds arriving off axis to the microphone.

Omnidirectional microphone: A microphone that is equally sensitive to sounds arriving from all directions.

On-location recording: A recording made outside the studio, in a room or hall where the music is normally performed or practiced.

O.R.T.F.: Abbreviation for Office de Radiodiffusion-Television Française, or French Broadcasting Network. A near-coincident stereo microphone technique in which two cardioid microphones are angled apart 110° and spaced 17 cm apart horizontally.

OSS system: Abbreviation for Optimal Stereo Signal system. See Jecklin disk.

Pan pot: Abbreviation for panoramic potentiometer. In each input module in a mixing console, a control that divides a signal between two channels in an adjustable ratio. By doing so, a pan pot controls the location of a sonic image between a stereo pair of loudspeakers.

Parabolic microphone: A highly directional microphone made of a parabola-shaped sound reflector which focuses sound into the microphone element.

Perspective: In the reproduction of a recording, the audible sense of distance to the musical ensemble, the point of view. A close perspective has a high ratio of direct sound to reverberant sound; a distant perspective has a low ratio of direct sound to reverberant sound.

Phantom image: See Image.

Phantom power: A DC voltage (usually 12 to 48 volts) applied to microphone signal conductors to power condenser microphones.

Phase: The degree of progression in the cycle of a wave, where one complete cycle is 360°.

Phase cancellation, phase interference: The cancellation of certain frequency components of a signal that occurs when the signal is combined with its delayed replica. At certain frequencies, the direct and delayed signals are of equal level and opposite polarity (180° out of phase), and when combined, they cancel out. The result is a comb-filter frequency response having a periodic series of peaks and dips. Phase interference can occur between the signals of two microphones picking up the same source at different distances, or can occur at a microphone picking up both a direct sound and its reflection from a nearby surface.

Phase shift: The difference in degrees of phase angle between corresponding points on two waves. If one wave is delayed with respect to another, there is a phase shift between them of $2\pi FT$, where $\pi = 3.14$, $F = $ frequency in Hz, and $T = $ delay in seconds.

Pinnae: The outer ears. Reflections from folds of skin in the pinnae aid in localizing sounds.

Polar pattern: The directional pickup pattern of a microphone; a graph of microphone sensitivity versus angle of sound incidence. Examples of polar patterns are omnidirectional, bidirectional, and unidirectional. Subsets of unidirectional are cardioid, supercardioid, and hypercardioid.

Polarity: Referring to the positive or negative direction of an electrical, acoustical, or magnetic force. Two identical signals in opposite polarity are 180° out-of-phase with each other at all frequencies.

Pressure Zone Microphone®: A boundary microphone constructed with the microphone diaphragm parallel with, and facing, a reflective surface.

Printed ribbon microphone: A dynamic microphone having a plastic diaphragm with an implanted ribbon.

Record: To store an event in permanent form; usually, to store an audio signal in magnetic form on magnetic tape.

Reflected sound: Sound waves that reach the listener after being reflected from one or more surfaces.

Remote recording: See On-location recording.

Reverberation: The persistence of sound in a room after the original sound has ceased, caused by multiple sound reflections (echoes), decreasing in intensity with time, so closely spaced in time as to merge into a single continuous sound, eventually being completely absorbed by the inner surfaces of the room. The timing of the echoes is random, and the echoes increase in number as they decay.

An example of reverberation is the sound you hear just after you shout in an empty gymnasium.

An echo is a discrete repetition of a sound, while reverberation is a continuous fade-out sound.

Artificial reverberation is reverberation in an audio signal created mechanically or electronically rather than acoustically.

Ribbon microphone: A dynamic microphone in which the conductor is a long metallic diaphragm (ribbon) suspended in a magnetic field. Usually a ribbon microphone has a bidirectional (figure-eight) polar pattern and can be used for the Blumlein method of stereo miking.

Sampling: Recording a short sound event into computer memory. The audio signal is converted into digital data representing the signal waveform, and the data is stored in memory chips for later playback.

SASS™: Abbreviation for Stereo Ambient Sampling System™. A stereo microphone using two boundary microphones, each on a 5-inch square panel, angled apart and ear-spaced, with a baffle between the microphones.

Semi-coincident: See Near-coincident.

Sensitivity: The output of a microphone in volts for a given input in sound pressure level.

Shock mount: A suspension system which mechanically isolates a microphone from its stand or boom, preventing the transfer of mechanical vibrations.

Shotgun microphone (line microphone): A highly directional microphone made of a slotted "line interference" tube mounted in front of a hypercardioid microphone capsule.

Shuffling: See Spatial equalization.

Signal: A varying electrical voltage that represents information, such as a sound wave.

Signal-to-noise ratio: The ratio in decibels between signal voltage and noise voltage. An audio component with a high signal-to-noise ratio has little background noise accompanying the signal; a component with a low signal-to-noise ratio is noisy.

Size: See Focus.

Sound: Longitudinal vibrations in a medium in the frequency range 20 Hz to 20,000 Hz. Also, the perception of such vibrations.

Sound pressure level (SPL): The acoustic pressure of a sound wave, measured in decibels above the threshold of hearing. dB SPL = 20 log (P/P_{ref}), where $P_{ref} = 0.0002$ dyne/cm^2.

Sound wave: The periodic variations in sound pressure radiating from a sound source.

Spaced pair: A stereo microphone technique using two identical microphones horizontally spaced several feet apart, usually aiming straight ahead toward the sound source. Time differences between channels localize the sound images.

Spaciousness: A sense of three dimensional space or air; a wide stereo spread. The ratio of L−R (difference) information to L+R (sum) information in the stereo signals. A spaciousness of 1 or greater is desirable.

Spatial equalization: A low-frequency shelving boost in the L−R (difference) signal of a stereo program, and a complementary shelving cut in the L+R (sum) signal, in order to align the locations of the low- and high-frequency components of images, and to increase spaciousness or stereo separation.

Speaker: See Loudspeaker.

Spectrum: Level versus frequency of a signal; the relative levels of fundamentals and harmonics.

SPL: See Sound pressure level.

Spot microphone: A close-placed microphone that is mixed with more-distant microphones to add presence or to improve the balance.

Stage width: See Stereo spread.

Stereo, stereophonic: An audio recording and reproduction system with correlated information between two (usually) channels, meant to be heard over two or more loudspeakers, to give the illusion of sound-source localization and depth. Stereo means "solid."

Stereo bar, stereo microphone adapter: A microphone stand adapter that mounts two microphones on a single stand for convenient stereo miking.

Stereo imaging: The ability of a stereo recording or reproduction system to form clearly defined audio images at various locations between, or outside of, a stereo pair of loudspeakers.

Stereo microphone: A microphone containing two microphone capsules in a single housing for convenient stereo recording. The capsules usually are coincident.

Stereo spread: The reproduced stage width. The distance between the reproduced images of the left side and right side of a musical ensemble.

Supercardioid microphone: A unidirectional microphone that attenuates side-arriving sounds by 8.7 dB, attenuates rear-arriving sounds by 11.4 dB, and has two nulls of maximum sound rejection at 125° off axis.

Surround sound: Stereo reproduction in which images surround the listener. The images of the musical instruments, and the hall reverberation, are heard in front between the speakers, and also to the sides and behind the listener.

Three-pin connector: A 3-pin professional audio connector for balanced signals. Pin 1 is connected to the cable shield; pin 2 is connected to the signal hot lead, and pin 3 connects to the signal return lead. Also see XLR-type connector.

Timbre: The subjective impression of spectrum and envelope. The quality of a sound that allows us to differentiate it from other sounds. For example, if you hear a trumpet, piano, and a drum, each has a different timbre or tone quality that identifies it as a particular instrument.

Tonal balance: The balance or volume relationships among different regions of the frequency spectrum, such as bass, mid-bass, mid-range, upper mid-range, and highs.

Track: A path on magnetic tape containing a single channel of audio.

Transaural stereo: A method of stereo recording for surround sound. During recording, the signals from a dummy head are processed for playback over loudspeakers, so that acoustic crosstalk around the head is cancelled. This crosstalk is the signal from the right speaker that reaches the left ear, and the signal from the left speaker that reaches the right ear. The net effect is to enable the listener to hear, over loudspeakers, what the dummy head heard in the original environment.

Transducer: A device that converts energy from one form to another, such as a microphone or loudspeaker.

Transformer: An electronic component containing two magnetically coupled coils of wire. The input signal is transferred magnetically to the output, without a direct connection between input and output.

Transient: A rapidly changing signal with a fast attack and a short decay, such as a drum beat.

Unidirectional microphone: A microphone that is most sensitive to sounds arriving from one direction—in front of the microphone. Examples are cardioid, supercardioid, and hypercardioid.

Virtual loudspeaker: A transaural image synthesized to simulate a loudspeaker placed at a desired location.

Wavelength: The physical length between corresponding points of successive waves. Low frequencies have long wavelengths; high frequencies have short wavelengths.

Windscreen: A fabric or foam-plastic screen that surrounds a microphone and reduces wind noise.

XLR-type connector: An ITT Cannon part number which has become the popular definition for a 3-pin professional audio connector. Also see Three-pin connector.

XY: See Coincident pair.

Index

Aachen Head, 99, 100, 157
   model HMS II dummy head, 153
A-B. *See* Spaced pair
Accent microphone. *See* Spot microphone
AKG Acoustics, Inc., 157
AKG C-34 stereo microphone, 146
AKG C-422 stereo microphone, 146, 147
AKG C-426B "Comb" stereo microphone, 147
AKG C-522 ENG stereo microphone, 146, 148
AKG H 10 stereo crossbar, 153
AKG H52 stereo suspension set, 154
AKG KM-235/1 stereo bar, 153
Ambience, 161
Ambience microphone, 161
Ambisonics control unit, 106
Amplitude, 43, 44, 161
AMS/Calrec soundfield microphone, 106, 147
AMS/Calrec ST250 stereo microphone, 147–148
AMS Industries Inc., 158
Angling, 48–49
Anti-crosstalk filters, 105
Archer International Developments, 106
Artificial head. *See* Dummy head
Association model, 38
Atal, B., 104
Audio Design, 106
Audio-Engineering Associates, 157
   MS38 dual-mode active MS matrix decoder, 70, 155
   MS380 TX matrix decoder, 155, 156

Audio Intervisual Design, 159
Audio-Technica, 158
   AT-825 XY stereo microphone, 148

Backplate, 5
Balance, 161
Bartlett, B., 38
B.A.S.E. (Bedini Audio Spacial Environment), 106
Bauck, J., 34, 35, 93, 104, 105
Bauer, B., 104
Bedini, John, 106
Bernfeld, Benjamin, 79
Beyerdynamic, 158
   MC742 stereo microphone, 148
Beyer ZMS1 mounting rail, 154
Bidirectional microphone, 1, 2, 20, 57–59, 161
Binaural recording, 35, 161
   and the artificial head, 96–101
Binaural reverberance suppression, 101
Binaural synthesis, 105
Biphonic processor, 104
Blauert, Jens, 38, 93
Blumlein, A., 50, 73
Blumlein technique, 19, 57–59, 68, 78, 79, 80, 161
Boers, P., 105
Booms, 11–12
Boundary microphones, 9–10, 83, 112, 161
   floor-mounted configured for MS, 87
   floor-mounted directional, 84–85
   floor-mounted spaced 4 ft apart, 84

173

174 ☐ STEREO MICROPHONE TECHNIQUES

L-squared array, 89, 90
L-squared floor array, 85–87
OSS floor array, 87
Pillon PZM® stereo shotgun, 89–90, 91
PZM® wedge, 88–89
raised-boundary methods, 87–94
SASS™, 90–94
Boyk, James, 95
Bray, W., 101
Bruck, Jerry, 87
Bruel & Kjaer, 158
   3529 and 3530 stereo microphone sets, 154
   4128 Head and Torso Simulator, 153

Cabot, R., 38
Calrec Audio Ltd., 158
Calrec Soundfield microphone, 73, 74, 106, 147
Camera Mart, 159
Capacitor microphone. *See* Condenser microphone
Cardioid microphone, 2, 56–62, 162
   angled 90° and spaced 30 cm, 78
   angled 110° and spaced 17 cm, 78
Ceoen, Carl, 60, 78–79
Channel, 162
Closely spaced. *See* Near-coincident
Coincident pair technique, 18–21, 112, 130, 142, 162
   Calrec Soundfield microphone, 73, 74
   coincident bidirectionals angled 90° apart, 57–59, 79
   coincident cardioids angled 90° apart, 57, 58, 78, 79
   coincident cardioids angled 120° to 135° apart, 57, 79
   coincident cardioids angled 135°, 78
   coincident cardioids angled 180° apart, 56
   coincident figure eights at 90°, 78
   coincident hypercardioids angled 120° apart, 79
   coincident systems with spatial equalization, 73, 75

features of, 24
hypercardioids angled 110° apart, 59–60
MS (mid-side), 67–73
   with spatial equalization, 73, 75
XY shotgun microphones, 60
Coles Biophonic Mr. Aural and Lady Aural dummy heads, 153
Comb-filter effect, 117, 162
Company addresses, 157–160
Compressor, 162
Condamines, R., 62
Condenser microphone, 5, 7, 112, 162
Cooper, D., 34, 35, 38, 56, 93, 100, 104–105
Crosstalk, 35, 162
Crosstalk canceller, 102, 103, 104–105
Crown International, 158
Crown SASS-B™ stereo microphone, 148
Crown SASS-P™ PZM® stereo microphone, 9, 148

Davis, Carolyn, 107
Davis, Don, 107
dB, 162
Dead, 162
Decibel, 162–163
Delay, 163
Depth, 29, 163
Diaphragm, 5
Diffuse field, 163
Diffuse field equalization, 100
Directional microphone, 3, 10, 18, 112, 163
Direct sound, 163
Distortion, 163
Documentaries, 142
Dooley, W., 72
"Double MS" technique, 72, 117, 118
Dry, 163
Dummy head(s), 96–97, 163
   equalization, 100
   imaging with loudspeakers, 100–101
   in-head localization, 99
   mechanics of, 97–99
   types of, 153

Dutton, G., 38
Dynamic microphone, 163
Dynamic range, 163

Eargle, John, 77, 117, 121
Echo, 163
Electret-condenser microphone, 163
Electronic news gathering (E.N.G.), 133–134, 135
Electrostatic field, 163
Elevation, 29, 163
Envelope, 163
Equalization (EQ), 164
 artificial-head, 100
Equalizer, 164
Equipment, 111–113
Exaggerated-separation effect, 22, 77

Faulkner, Tony, 76
Faulkner phased-array system, 76, 164
Feature films, 141–142
Focus, 164
Fostex Corporation of America, 158
Fostex M20RP mid-side stereo microphone, 148
Fostex M22RP mid-side stereo microphone, 148
Free field, 9, 164
Free-field equalization, 100
Frequency, 164
Frequency response, 164
Frontal free-field equalization, 100
Fundamental, 164
Fusion, 27, 164

Gamma Electronics, 106
Generation, 164
Generation loss, 164
Genuit, K., 100, 101, 105
Gerlach, Henning, 60, 72, 73
Gerzon, Michael, 34, 49, 56, 57
Gibson, Dan, 143
Gierlich, H. W., 100, 105
Gotham Audio Corp., 157, 158, 159
Griesinger, D., 38, 49, 50, 63, 72, 73, 75, 93, 100, 101, 106

Harmonic, 164
Head-diffraction filters, 105
Headphones, 164
Hertz (Hz), 164
Hibbing, M., 80
Hiss, 164
Hole-in-the-middle, 22, 77, 126
Holophonic Systems, 158
 H-1 Holophone, 153
Huggonet, C., 80
Hypercardioid microphone, 2, 3, 59–60, 164

Image, 165
Image focus, 29
Image movement, 29
Imaging, 17, 56, 165. See also Stereo imaging
 artificial head, with loudspeakers, 100–101
Industrial productions, 142
Input module, 165
Intensity stereo (XY stereo), 165
Interaural differences, 33, 34, 35, 38
Interchannel differences, 33, 34, 35, 36–37, 38
Iovine, Jimmy, 106
ITE/PAR, 107, 165

Jecklin disk, 63–64, 148, 165
Josephson Engineering, 87, 159
 C-622 OSS-II stereo microphone, 148–149
 OSS-disk, 148
 SCH-disk, 149
 Stereo-Q, 149
Jouhaneau, J., 80
Julstrom, Steve, 106
Junction box, 13
JVC, 104

KEMAR dummy head, 100, 153
Kilo, 165
Klayman, Arnold, 106
Knowles Electronics, 153
Koshigoe, S., 34
Kugelflachenmikrofon, 99

Lamm, Mike, 85, 89
Lees, John, 106
Lehman, John, 85, 89
Level, 165
Lexicon CP-1 digital audio environment processor, 105–106
Live, 165
Live recording, 165
Localization, 16, 29, 165
  by amplitude differences, 39–41
  by amplitude and time differences, 43
  of images between speakers, 32–34
  in-head, 99
  of real sound sources, 29–32
  by time differences, 41
Localization accuracy, 53–55, 165
Location, 27, 28, 165
Loudspeaker, 165
Lowe, Dan, 106

Matrix decoders, 155–157
Mic, 165, 166
Mic level, 165
Microphone, 166
Microphone techniques, 166
Mid-side (MS) technique, 19–20, 60, 67–68, 78, 79, 80, 87, 166
  advantages of, 69–72
  comparison with XY coincident methods, 80
  disadvantages of, 72
  double, 72, 117, 118
  matrix box, 68–69
  with mid shotgun, 72–73
Mix, 166
Mixer, 166
Monitor, 116, 132–133, 166
Monaural, 166
Mono, monophonic, 166
Mono-compatible, 132, 166
Mounting hardware, 25
Moving-coil microphone, 5, 6, 7
MS Polar Pattern Program, The, 157
MS recording. *See* Mid-side, 166
Muddy, 166
Murphy, Shawn, 136
Myers, Pete, 106

Nakabayashi, K., 34
Natural image, 29
Near-coincident techniques, 23–24, 45, 112, 130, 142, 143, 166
  Faulkner phased-array system, 76
  features of, 24
  NOS system, 62
  ORTF system, 60–62
  OSS system, 63–64
  spaced-pair hybrid, 77
  with spatial equalization, 76
  Stereo 180 system, 75–76
Near-field monitoring, 166
Neumann Microphones, 137, 159
  DS21 dual microphone mount, 154
  KU 81i dummy head, 100, 153, 154
  RSM 190S system, 149, 150
  RSM 191S system, 149
  SM69 fet stereo microphone, 149
  USM 69i stereo microphone, 10, 149
  Z 240 matrix decoder, 156
Noise, 166
NOS system, 80, 143, 166
  cardioids angled 90° apart and spaced 30 cm horizontally, 62, 78, 79

Off axis, 166
Off-axis coloration, 167
Olson, Lynn T., 75
Omnidirectional microphones, 1, 3, 10, 23, 112, 167
  spaced 3 ft apart, 65
  spaced 5 ft apart, 66–67
  spaced 9.5 ft apart, 79
  spaced 10 ft apart, 66
  spaced 50 cm, 78
On-location recording, 167
Optimal stereo signal. *See* OSS
ORTF, 23–24, 79, 80, 143, 167
  cardioids angled 110° apart and spaced 17 cm horizontally, 60–62, 78, 79
OSS system, 63–64, 130, 143, 167
  boundary microphone floor array, 87
Output level, 8

Panasonic SY-DS1 Surround Sound Processor, 106
Pan pot, 105, 167
Parabolic microphone, 11, 167
Parades, stereo television of, 136–138
Pearl Mikrofonlab, 159
   MS-2 stereo microphone, 149
   MS-9 stereo microphone, 149
   TL-4 stereo microphone, 149
Perspective, 167
Phantom image, 41, 43, 167
Phantom power, 13, 167
Phase, 167
Phase cancellation, 167
Phase interference, 167
Phase shift, 167–168
Pillon, Gary, 89, 142
Pillon PZM® stereo shotgun, 89–90, 91
Ping pong effect, 22
Pinnae, 168
Pizzi, Skip, 68, 72, 117, 138
Plays, radio, 140–141
Polar patterns, 1–3, 44, 168
   advantages of, 3
   illustration of various, 4
   other considerations, 3
   transducer type, 5–7
Polarity, 33–34, 168
Pop music, stereo miking for, 122–123
Posthorn Recordings, 87
Pressure Zone Microphone®, 168
Printed ribbon microphone, 168
PZM® microphones, 85, 130, 137, 142
PZM® wedge, 88–89

Q-Biphonic, 104
Q Sound, 106
Quasi-coincident. *See* Near-coincident.

Radio. *See* Stereo radio
Radio Shack (Realistic) 33-1065 stereo microphone, 150
Radio Shack (Realistic) 33-2012 stereo microphone, 150
Record, 168

Recording angle, 77
Recording procedures
   choosing the recording site, 113
   equipment, 111–113
   miking distance, 116–117
   monitoring, 116
   multitrack recording, 122
   session setup, 113–116
   setting levels, 122
   soloist pickup and spot microphones, 120–122
   stereo miking for pop music, 122–123
   stereo-spread control, 117–119
   troubleshooting stereo sound, 124–128
Reflected sound, 168
Reverberation, 168
Ribbon microphone, 5, 6, 7, 168
Rumsey, F., 38
Ryckman, Lawrence G., 106

Salava, T., 104
Sampling, 143–144, 168
Sanken Microphones, 159
   CMS 2 stereo microphone, 150, 151
   CMS-7 stereo microphone, 150, 152
   CMS-9 stereo microphone, 150
   MBB-II matrix box, 156, 157
   M-5 matrix box, 156
   XY stereo microphone mount, 154
SASS™, 90, 143, 168–169
   advantages and disadvantages of, 93–94
   construction of, 91
   frequency response of, 91–93
   localization mechanisms of, 93
Schaefer, Ralph, 106
Schneider disk, 149
Schoeps Microphones, 159
   CMTS 501U stereo microphone, 151, 152
   MSTC 34, 44, and 54 stereo microphone, 150
   STC ORTF system stereo mounting bar, 155

178 □ STEREO MICROPHONE TECHNIQUES

UMS 20 Universal stereo bracket, 12, 155
UMSC Universal Collette stereo bracket, 155
VMS 52 UB matrix decoder, 156
Schroeder, M., 104
Self-noise, 8, 112
Semi-coincident. *See* Near-coincident
Sennheiser Electronic Corporation, 160
   MKE 2002 Set, 153
   MZT 235 and MZS 235 adapter bars, 155
Sensitivity, 7–8, 169
Session setup, 113–116
Shock mount, 12, 112, 114, 169
Shotgun microphone (line microphone), 11, 60, 72–73, 137, 142, 169
Shuffling. *See* Spatial equalization
Shure Brothers Inc., 160
   Model A27M stereo microphone adapter, 155, 156
   Surround Sound Processor, 106
   VP88 stereo microphone, 151
Signal, 169
Signal-to-noise ratio, 8, 169
Size, 169
Smith, Bennett, 79
Snake box, 13, 115
Soloist pickup microphones, 120–122
Sony Communications Products Co., 160
   CM-MS5 stereo microphone, 151
   ECM-979 stereo microphone, 151
Sound, 169
Sound effects, 142–143
Sound pressure level (SPL), 7, 169
Sound Retrieval System, 106
Sound wave, 169
Spaced-pair techniques, 21–23, 64–65, 112, 169
   features of, 24
   near-coincident hybrid, 77
   omnis spaced 3 ft apart, 65
   omnis spaced 5 ft apart, 66–67
   omnis spaced 10 ft apart, 66
Spacing, 48–49

Spaciousness, 49–50, 169
Spatial equalization, 35, 49–50, 73, 75, 101, 126, 169
Speaker. *See* Loudspeaker
Spectrum, 169
SPL. *See* Sound pressure level
Splitter, 13
Sporting events, stereo televising of, 138–139
Spot microphones, 120–122, 170
Stage width. *See* Stereo spread
Standard deviation, 77
Stands, 11–12
Stereo (stereophonic), xiii, 14, 15, 170
Stereo Ambient Sampling System™. *See* SASS™
Stereo bar(s), 12, 25, 112, 114, 170
   types of, 153–155
Stereo imaging, 170
   choosing angling and spacing, 48–49
   currently used image-localization mechanisms, 38–44
   definitions related to, 27–29
   localizing images between speakers, 32–34
   localizing real sound sources, 29–32
   predicting image locations, 44–48
   requirements for natural imaging over loudspeakers, 34–38
   spaciousness and spatial equalization, 49–50
Stereo microphone adapter(s). *See* Stereo bar(s).
Stereo microphone techniques, 10, 14, 15, 170
   advantages of using, 14–15
   applications for, 15
   comparison of techniques, 24
   goals of, 16–18
   microphone requirements, 25
   miking distance, 116–117
   mounting hardware, 25
   stereo-spread control, 117–119
   types of, 18–24, 146–152
Stereo 180 system, 75–76

Index □ **179**

Stereo radio
  for group discussions, 139–140
  for plays, 140–141
Stereosonic technique, 57–59
Stereo spread, 27–28, 170
Stereo television
  audience reaction and, 134–136
  electronic news gathering (E.N.G.), 133–134, 135
  imaging considerations of, 131
  monitoring of, 132–133
  mono-compatibility of, 132
  of parades, 136–138
  of sporting events, 138–139
Streicher, Ron, 72, 137
Supercardioid microphone, 2, 3, 137, 170
Surround sound, 105, 170
Synclavier systems, 143
Synergetic Audio Concepts, 107

Telarc Records, 66
Television. *See* Stereo television
Theile, Gunther, 35, 38, 93, 99, 100
3-D Audio, 106
Three-pin connector, 170
Timbre, 170
Time difference, 33, 41, 43, 45
Tonal balance, 170
Track, 170
Transaural converter, 35
Transaural stereo, 35, 170–171
  Cooper and Bauck's crosstalk canceller, 104–105
  history of, 104
  Lexicon's transaural processor, 105–106
  mechanics of, 101–104
  other surround-sound systems, 106–107
Transducer, 5–7, 171

Transformer, 171
Transient, 171
Troubleshooting
  bad balance, 127
  bad tonal balance, 128
  distortion in the microphone signal, 124
  early reflections too loud, 127
  excessive separation or hole-in-the-middle, 126
  images shifted to one side, 126
  insufficient ambience, hall reverberation, or room acoustics, 124
  lacks depth, 127
  lacks spaciousness, 127
  muddy bass, 127–128
  narrow stereo spread, 125
  poorly focused images, 126
  rumble from air conditioning, trucks, and so on, 128
  too detailed, too close, too edgy, 124–125
  too distant, 125

Unidirectional microphone, 1, 2–3, 10, 171

Vanderlyn, P., 49, 73
Viale Parioli, 159
Virtual loudspeakers, 105, 171

Wavelength, 171
Williams, Michael, 38, 77–78
Windscreen, 133, 143
Woram, John, 157

XLR-type connector, 171
XY. *See* Coincident pair

Yakus, Shelly, 106